CUBED

ERNŐ RUBIK

CUBED

THE PUZZLE OF US ALL

FLATIRON
BOOKS
NEW YORK

www.flatironbooks.com

Designed by Jonathan Bennett

Library of Congress Cataloging-in-Publication Data

Names: Rubik, Ernő, author.
Title: Cubed : the puzzle of us all / Ernő Rubik.
Description: First edition. | New York : Flatiron Books, 2020.
Identifiers: LCCN 2020019368 | ISBN 9781250217776 (hardcover) |
    ISBN 9781250217783 (ebook)
Subjects: LCSH: Rubik, Ernő. | Rubik's Cube. | Inventors—
    Hungary—Biography.
Classification: LCC QA491 .R83 2020 | DDC 793.74 [B] —dc23
LC record available at https://lccn.loc.gov/2020019368

Our books may be purchased in bulk for promotional, educational, or business use. Please contact your local bookseller or the Macmillan Corporate and Premium Sales Department at 1-800-221-7945, extension 5442, or by email at MacmillanSpecialMarkets@macmillan.com.

First Edition: 2020

10   9   8   7   6   5   4   3   2   1

To Ágnes

CONTENTS

If at first, the idea is not absurd,
there is no hope for it.

—ALBERT EINSTEIN

CUBED

INTRODUCTION

My official name is the Rubik's Cube. "Cube Rubik" sounds more natural to me, but nobody has really asked me about my feelings. If I were of noble blood, you could call me the "Hungarian Magic Cube von Rubik," but I'm not. Personally, I prefer "Magic Cube" because it reminds me of my childhood, but my friends just call me "the Cube," and you may call me that as well. We probably have met already, since I've traveled all over the world and many millions have touched me and been touched by me over the decades. Even if you weren't one of them, please don't worry. (I never worry, by the way.)

You've probably seen me in the hands of people, or my image sometime, somewhere: on TV screens, T-shirts, magazine covers; in movies, YouTube videos, books; as part of tattoos, sculptures, album art; maybe in school . . . and I could go on and on. They say that one in every seven people in the world today has played with me! That is more than a billion. Can you imagine?

Even though you certainly have seen me, it must be strange to actually hear from me, so let me explain. You are reading a book by Rubik, the person who gave me life in 1974. There is nothing conventional about this book—especially the man who wrote it (he believes the contrary)—and it became clear while this was going on that I needed to be included. I wanted to help him tell the story, because I'm its most authentic witness! (He hates to write and has a poor memory.) And since every puzzle has rules, here are mine: I can't think, but I can express myself. I can't read or write, but I hear a lot and never forget. I am very simple/complex. I am colorful and happy. I met a young Hungarian fellow a long time ago (and now we are not so young . . .) and since then, we've been a team.

Teamwork has been my life. If you've ever picked me up and played with me, you and I formed a team. Now that you are reading, we are another team; you the Reader, and me with Rubik, the Writers. A group of three. As a 3x3x3, I think that the number three is magical. It has such perfect symmetries.

If all this seems bizarre to you, simply relax and open your mind. As Albert Einstein said, "The true sign of intelligence is not knowledge but imagination."

So let's play!

—*The Cube*

ERNŐ RUBIK

Who in the world am I?
Ah, that's the great puzzle.

—LEWIS CARROLL

I GUESS MANY PARENTS have had the same experience I have had: Suddenly observing their own children with a moment of curious detachment and wonder, and not at all from the perspective of being a mother or a father. In these revealing and sometimes beautiful moments that I have had with my children, it is as if I were meeting them for the first time, and I see them being deeply involved in a world that has nothing to do with me. When that happens, and it is never planned and does not occur often, I am startled to see in them qualities that I have never appreciated before. A tone of voice, perhaps, or a way of thinking that is totally unpredictable, surprising, or maybe even the sudden revelation of a strange interest or a curious hobby I had never suspected they had.

It has been the same with my eldest child: the Cube. There

are some languages that have genders, and in these languages the word "cube" is almost always masculine—*le cube* in French, or *der Würfel* in German, for example—so when I refer to the Cube, I will use that distinction. He is my boy, my son. If you take a ball in your hand, it is a totally different feeling: soft, supple—a cube is a boy with edges and muscles.

Even as much as he has defined my life for nearly half a century, I can still be caught off guard by discovering some unexpected quality or character in him. Sometimes it is as simple as when I am playing with the rigid plastic pieces, but I am struck again and again by how they behave. The interplay of forces, the cohesive strength of all the elements, remind me of a drop of water floating weightlessly on a table, contained into a spherical shape by surface tension. I like the possibilities the Cube contains, and simply adore the visual pleasure of its shape. Often, the cubical shape is associated with an item that we have no control over, like dice. But there is nothing haphazard or out of control with the Cube. That is, as long as you are willing to give it some patience and some curiosity.

I HATE TO WRITE. Yet here I am, writing this book. There is no way back. Writing as an exercise is both technical and intellectual. Maybe being left-handed added some awkwardness to learning to write in a right-handed world. In retrospect, I was fortunate to have a teacher who did not force children to go against their natural proclivities. There was no pressure at all beyond the encouragement that I do the required work. My more pressing question with writing is abstract: How can we possibly capture in words all the dimensions of our lives?

That is not to say that I am not an avid reader. But when the writing involves a life—specifically *my* life—I find the medium almost paralyzing. This is not the first time I have confronted the challenge of writing about my experiences, my time with the Cube, and, inevitably, my life story. So far I have easily yielded to the temptation of not writing at all. But there is also the equally strong temptation of doing something well, of attempting to do something that feels authentic. Finally, I decided to approach the task of writing as if it were a puzzle, and I considered the model that I know best of all: the Cube, which I discovered in 1974. As an object, it shares many characteristics with the kind of writing I like best. It is simple and complex; it has movement and stability. There is what we see, and then there is its hidden structure.

Simple and complex. Moving and stable. Hidden and exposed. I believe contradictions are not opposites to be resolved, but counterpoints to be embraced. Rather than becoming frustrated by what seems irreconcilable in a contradiction, the better option is to appreciate that a contradiction helps us make connections we may never have considered. One can never fully capture three dimensions on a page. And yet, framing the many themes in my work and in my life in terms of contradictions could add dimensions that may make it easier for me to write.

IT PROBABLY GOES without saying that the Cube has attracted more attention than I could have ever imagined. It is a curious fact—one that surprises me as much as anyone— that for so many decades, during a time of an unprecedented technological revolution, fascination with such a simple, "low-tech" object has survived. And, in fact, this fascination has evolved. The Cube has been a toy for children, an intensely competitive sport, and a vehicle for high-tech explorations and discoveries in artificial intelligence and bewildering mathematics. Blame has been cast on the Cube for divorces (and marriages), and for ailments known as "the cubist's thumb" and "Rubik's wrist."

With all this attention has come . . . questions. Journalists, fans of the Cube, or casual acquaintances around the world, often ask me the same questions, as if I could easily provide answers that would reveal all the mysteries of my puzzle. They have hardly changed over the years, so let's dispense with them at the outset, shall we?

> Q: How did you invent the Cube?
>
> A: I sat down to think about a geometrical problem and how to illustrate it. I made something that became the Cube.
>
> Q: How long did it take?
>
> A: I began in the spring of 1974 and applied for a patent the following January.
>
> Q: What is your record for solving it?
>
> A: I have no idea. I have never measured my time.
>
> Q: What are the tricks?
>
> A: There are no tricks. At all.
>
> Q: Why did you invent the Cube? [For me this is the most irritating question.]
>
> A: I found a problem that captured my imagination and did not let me escape.

If these are the questions that a reader expects to be answered in this book, those are the responses, and one can stop reading right away. At the same time, I'm aware that asking a true question is more difficult than answering one. In the end, revealing or interesting answers can be given only in response to good questions.

What, then, are the questions that I would prefer to be asked? Well, one that may have already occurred to you is this: After all these years of "hating to write," why did I decide to write a book? I must admit, my motives were rather selfish. For all its shortcomings, writing does offer a chance to explore some questions in order to gain a deeper understanding. So even though I may hate to write, I am always eager to try to understand things better, especially those things that we take for granted. What makes us tick? What makes us create? And how are people inspired to make something that has never been made before?

This is also my attempt to try to more completely understand the remarkable popularity and endurance of the Cube. What does it say about the ways our mind works? Does it suggest there are certain universal qualities that bring us together?

One example of the Cube's ability to bridge seemingly unbridgeable differences occurred very early on. In 1978, one year after it had appeared for the first time in toy shops in my home city of Budapest, I took my newborn baby daughter to a playground.

And there was my Cube! In fact, there were two Cubes in the park, and two very different people playing with them! The first was a little boy who was about eight years old. Quite content and extremely dirty, he sat on the ground, playing with the Cube—a small Oliver Twist, twisting it. The second Cube emerged from the elegant handbag of a youthful mother in her thirties who must just have come from the beauty salon. She was sitting on a bench and cast only an occasional glance at her baby in its stroller, so thoroughly was she immersed in tackling the Cube. It was astounding to see on the faces of these completely opposite people the very same expression.

Since then, I have seen that expression on faces all over the world. They are faces in repose but also intently engaged. Concentrating, turning inward, losing touch with their surroundings and the external world. They look as if they are in a state of meditation, except instead of being lost internally, they are engaged and active. They are suspended within a rare moment of peaceful coexistence between order and chaos.

I REALIZED I'VE TAKEN something for granted: Just like I hate to write but am still writing a book, you may not like to read but are reading one. If so, thank you for taking a look at my book anyway. You don't have to read it all in one sitting, or from the first page to the last. You are free to discover it as you want, and my hope is that you will permit yourself to get a little bit lost. In these pages, some of the puzzle pieces of my thoughts, insights, and observations may appear to be scrambled. Like the Cube, the internal structure is hidden, and what ultimately happens depends on you. Because every reader is different—bringing their own interests, talents, dreams, professions, passions, and contradictions to this or any book—there is not a single "right" way to read. All the pieces contained here may not fall into obvious places, and they don't need to.

This book will touch upon many things: creativity, symmetry, education, architecture, questions, playfulness, contradictions, beauty. But at its core, this book is about puzzles. It is about the puzzle of myself. It is about the puzzle of this strange object I discovered almost fifty years ago. And it is about the puzzle of us all.

MY FATHER was not a playful man. Ernő Rubik Sr. was once a well-known name in the field of aviation—and not only in Hungary. He was obsessed with creating the perfect glider. He had several patents, and designed more than thirty airplane and glider models and also a mini car made of aluminum. But only when I was an adult did I realize that every time he figured out the structure and the materials and all the details of his designs, he was solving very practical and complicated puzzles. Perhaps I saw him working on his plans and was inspired, or maybe I was just a curious little boy, but from the time I was a small kid in Budapest, I sought out puzzles and would spend hours immersed in their challenges. One of my favorite things to do was to devise strategies for new and more efficient solutions.

I liked different puzzles for different reasons and their different capacities. I liked some because of their flexibility and capacity for change. I liked others because their ideas were expressed with such simplicity. I liked still others because they provided the framework for improvisation. I liked difficult puzzles more than easy ones. I remember the curiosity, focus, periods of disorientation and frustration, some excitement when crucial connections were made, and then the sense of accomplishment when arriving at the solution.

Interest in puzzles is nearly universal. They have been around for much of human history. Anthropologists digging up pieces of the past and piecing them together discover puzzles all over the world. What I found in 1974 emerged from an entire lineage of puzzles that have inspired and baffled players since ancient times.

PLAYING WITH PUZZLES when I was a child trained my mind. I became familiar with the nature of their questions and answering them. I was not assigned these puzzles, was not graded on my performance, nor was anyone observing whether I solved them or not. If I failed or had trouble with one, I could start again on it the next day. This entertainment was solitary. Without an opponent, I was always the winner—not that I really thought that way. What most captured me was that I could use these puzzles as a starting point to discover something else.

Puzzles bring out important qualities in each of us: concentration, curiosity, a sense of play, the eagerness to discover a solution. These are the very same qualities that form the bedrock for all human creativity. Puzzles are not just entertainment or devices for killing time. For us, as for our ancestors, they help point the way to our creative potential. *If you are curious, you will find the puzzles around you. If you are determined, you will solve them.*

One that I played with very early on was the tangram, a deceptively simple geometric puzzle that is, in my view, not really a puzzle because it does not set a well-defined task. Originating in ancient China, a tangram is a square sliced up into seven pieces, or "tans": five triangles of varying sizes, a parallelogram, and a square. The challenge is to fashion from these simple elements a variety of unique figures. Sometimes, one can fit them all into a square. Other times, one may feel more whimsical and create figures from them. Usually it is an accidental composition of elements.

ERNŐ RUBIK

You can't have a theory in a mathematical sense to solve a tangram, or to say why these contours look like a man, the other looks like a tiger, and the third looks like a flower. You cannot imagine a simpler game, and yet from these pieces, an endless number of interesting figures can be constructed. The tangram appealed to me because it was very free. In a sense, it is close to art, since depending on how the pieces are assembled and the attitude one has when manipulating them, one can create very artistic results. I was one of those children who spent hours drawing and painting. Drawing something when I sat in class was a fine distraction when there were some subjects (or teachers) who bored me. With the tangram, sometimes I would draw on the pieces themselves so that when they were put together, they created something abstract and beautiful.

WHEN I WAS ABOUT FIVE OR SIX, I received a 15 puzzle as a present. I think that its original intent was to keep me occupied for the few hours on the train that it took to get from Budapest to Lake Balaton. Over the years, my father built us a cottage and we would spend the summer there. The original 15 puzzle was a flat box with fifteen squares that were numbered 1 to 15 and fitted into a four-by-four grid. There was always one empty space, which gave you the potential to move the pieces by sliding them.

In general, the challenge is to see how many possibilities, how many permutations or combinations of elements you can come up with. Another challenge was to see how many different ways, or in how many different permutations, the elements numbered from 1 to 15 could be arranged in the grids without taking them out and putting them back together. You have to follow the rule of sliding the pieces by filling the empty square. In that way it is a closed system. Today you can buy versions that are made from plastic with tongue-and-groove connections between pieces so you can't take them out of the frame. I prefer the old one that I had then. I could dump the pieces out from the box and put them back, scrambled. I especially liked the metallic sound when I played with it.

When the elements were inserted randomly, you needed to arrange the sequences by sliding the pieces. As a process, it was very simple. It was not a question of complexity but rather one of order and rules. If you have sequences of numbers in which each is valued as one, with no single one that is equal, they then can be arranged from the lowest to the highest. A simple law showed if something was possible and if something wasn't. You found the solution by discovering that not the individual pieces but the movements of the whole were important. If my parents hoped that this would keep me occupied for the few hours on the train, they must have been disappointed. I managed to solve it quickly.

There is no doubt that I learned from classics like the tangram and the 15 puzzle—but the pentomino was even more significant for me. "Pentomino," a term invented by American mathematician Solomon W. Golomb, means a shape consisting of five squares joined together by their sides. There are twelve different ways to arrange five squares. What is the task? The basic goal is to fill in rectangles; you can have different ones depending on the size. As one element consists of five squares, the area of the twelve different pentominoes is sixty squares (because $60 = 3 \times 4 \times 5$, so you can fill the 3x20, the 4x15, the 5x12, or 6x10 rectangles with the set and you can have more than one solution of each). Or you can create other things. You can fill the big 8x8 square with four empty small

ERNŐ RUBIK

squares in the middle or at the corners of the big one, or many different kinds of figures, and all of them are new tasks to solve.

Filling a surface with elements has so much potential and so many challenges. Mathematicians call it "tiling," which means covering the surface with elements that aren't overlapping. An enduring challenge that can feel unsolvable is to fill a rectangle with different sizes of squares. It becomes a very difficult task to create a "simple perfect squared square."

THE PENTOMINO WAS MY FIRST INTRODUCTION to recreational math and solving interesting geometrical problems. Geometry is very heuristic, very visual. For me, the visual aspect of the world was and remains the most important, the most formative experience.

The pentomino also offered further possibilities: You can make a three-dimensional version using cubes, not squares. It is called pentacubes and reveals how one can use cubes as the building blocks for more complex structures or designs. One of the basic options would be to choose one element from the twelve and double or triple its size in comparison to the rest. Another nice task is to fill a 3x4x5 box to store them.

In this early puzzle, I explored how cubes that are connected can be put together in a number of different ways. The visual capacity of the puzzle was beautiful.

I WAS NOT THE FIRST PERSON, obviously, to imagine the rich potential of the cube form. There are two predecessors that stand out to me. The Soma cube was created by a Danish scientist and poet named Piet Hein. He was a hero of World War II because of his work as a member of the Danish resistance, and then he went on to live a long life as a writer but also as an inventor of puzzles. I consider Hein's invention, like many puzzles, to be a work of art, especially if one considers how he framed his perception of art as "solving problems that cannot be formulated before they have been solved. The shaping of the question is part of the answer."

The Soma cube is closely related to the three-dimensional version of pentominoes. In this case there are seven pieces, six composed of four small cubes each, and one of three. But they are all different shapes: some are rectangular, some are L-shaped. The small cubes are joined to each other face-to-face. With these seven pieces, a 3x3x3 cube can be assembled. The Soma has 1,105,920 solutions.

The fact that the seventh piece is made of three small cubes—rather than four, like the others—means, in my opinion, the game lacks homogeneity. As a puzzle, it is a three-dimensional form, filling a 3x3x3 space. It looks like a cube; you can make it for yourself. The Soma cube is not an open puzzle like the tangram or the pentomino. They both have sets, and you can create your own challenge. The Soma cube is a classical puzzle whose challenge is to figure out the task that was determined by the puzzle's creator. It is a three-dimensional challenge.

I created my own version of it long before I even imagined making the Cube by trying to put together a 3x3x3 cube by using only elements that contained three equally small cubes. I created nine elements in which the number of small cubes was identical but the way they were joined was different. I used all of the potential combinations to join the three cubies, touching each other on faces and/or on edges. There are two elements that are joined only by faces. There are five elements that are joined only by edges. And two have both types of connections. There are 880 different solutions for the puzzle.

ERNŐ RUBIK

(This variation went to the market as Rubik's Bricks around 1990.)

The other important Cube predecessor for me is known as MacMahon's cube, which is also made up of cubes, much like a child's colored building blocks, in which all the faces are different colors and none are repeated. But the arrangement of the colors is different, and there are thirty different ways to make a cube with six colors. It is not as well-known as the others, but still it offers an interesting mathematical problem. There are thirty cubes whose faces have six colors, in all the possible permutations. The basic exercise is to choose one cube and then use eight others to make a 2x2x2 cube that has the same arrangement of colors as the first, with each face a single color and the interior faces matching. The biggest size that can be created by keeping the same rule is the 3x3x3 cube. From the perspective of the combinatorial question, there are thirty possible ways to arrange the colors on the six faces of the cube.

There are obvious similarities to the Cube, but with a very important difference: these cubies are all separated. Their elements are not connected physically. Once again these are combinatorial problems, which is to say, a challenge is to figure out how many different ways you can put them together. The nature of the task involves some kind of pattern recognition and imagination in which you need to find the right pieces and put them together.

In a strange way, sometimes one becomes a forerunner of one's forerunners.

What I mean is that we sometimes interpret an antecedent as if it were the consequence of something that occurred later. It is so very human.

There is a funny saying that has been ascribed to a Hungarian composer: "Schubert learned a lot from Schönberg."

Nowadays, if one sees an older puzzle or some geometric problems that resemble the Cube, the one thought that pops up is: Why didn't the inventor make the simple little leap to create the Rubik's Cube?

Not long ago, I thought of a new puzzle, this one involving twenty-seven small cubes that were not attached to each other. I used three colors for a set and tried to see if I could assemble a mono-color 3x3x3 cube with any of the colors. As it turned out, to solve it was much easier than it was to find the system of coloring. The main question was: How should the twenty-seven cubes be colored in so that they could be assembled in three different ways to see only one of the colors from the outside, so that at the same time the sides that touched were the same color? Finally I found the solution, not only for the number 3 but for the n.

ORSON WELLES ONCE APPEARED on a radio program and said, "Good evening, ladies and gentlemen, my name is Orson Welles. I am an actor. I am a writer. I am a producer. I am a director. I am a magician. I appear onstage and on the radio. Why are there so many of me and so few of you?" I adored the way he put it because I understood perfectly what he meant. There are so many of me because I am all the identities I carry with me all the time. All these definitions are restrictive, like so many different cells in a prison. All of us play many roles depending on the situation; like actors, we become the characters we are assigned. That's why it is so difficult to name the definitive one.

Sometimes I have appeared on television, and I am asked to introduce myself. The unspoken implication of the question for me is: Who *are* you? My answer is not very satisfying. "I am Ernő Rubik," I say, adding, "and I made the Cube."

It is a simple statement but doesn't really answer the question.

WHO AM I? There are so many possibilities: inventor, professor, architect, designer, sculptor, lecturer, editor, husband, father, grandfather, businessman, manager, writer (why not?), and so on. . . . How do I choose? I could say that I am all of these things, all at once, all the time, but with a different emphasis according to the situation, the task, or the activity.

There is a much longer list of who I am not.

I am not really the subject of this book. I am not a professional in any field. I am not really a writer. I am not a businessman. I am not young, but I don't feel old. I am not a carpenter, but I can make furniture. I am not in the navy, but I can sail a boat. I am not a gardener, but I love gardening. I could go on. I am an amateur at everything, including being an inventor. No one taught me how to learn, especially not my teachers.

When I think about the one aspect of my many identities that connects them, what I always come back to is that I am a playful man or, rather, a man who likes playing—what the Dutch scholar Johan Huizinga once called the *Homo ludens*.

Children are masters of play. It is often described as their most important job, and a basic part of how they learn. Children create rules when they are on their own and are very strict about following them. ("You are the doctor, I am the patient.") When they are playing games of their own invention, they will usually have extremely sophisticated guidelines that only a professional in the game could understand and follow. The older children get, the more complex the rules become, even as they are at the same time expressions of greater and greater imaginative freedom.

And then there is a turning point, when playful expressions of imagination get replaced by games imposed from the outside, with generally understood rules. By the time we are adults, the instinct for spontaneous play seems to have disappeared, and we seem eager to have rules constrain and define our actions.

All the exciting, imaginative play from childhood is gradually replaced by the more structured and conventional "play" of board games and team sports, in which there are clear winners and losers. Competition adds a level of discipline and the motivation to acquire greater expertise, and the individual performance is judged and put in a hierarchy of excellence. Unfortunately, the competitive spirit seems to replace the imaginative one. (Not that there is anything wrong with competition. My wife often complains about how competitive I get when we play Scrabble.)

In middle school, there was a brief period when I played chess. I found some real zealots to be my partners, and we played during classes and breaks, often "blind," i.e. without a board. In time, however, this passion shifted toward chess problem-solving, which suited my temperament better. I very much liked chess, not the game as much as the problems created by it. Using the chessboard, I would create new puzzles or solve others that had already been created. One challenge that I especially liked was called the knight's tour. It involved going through every square on the board with one knight never touching the same square twice until you ended up in the same place.

Like my favorite puzzles, the knight's tour is one I could play for hours, drawing the patterns on a chessboard matrix and seeing them emerge while I moved the knight, two squares in one direction and then one in another direction to finally arrive back at the starting point. Patterns emerged with symmetry and the richness of symmetries, like snowflakes. (This interest lasted for quite a long time. I used to solve the chess problems in the Hungarian chess periodical, and the first time I ever saw my name appear in print was in the list of the successful solvers.)

Too often as adults we seem to believe play is just a diversion, or another form of competition outside of the workplace. But play is one of the most serious things in the world. We often do things really well *only* when we do them playfully. We are more relaxed about them; the task becomes not a burden or a test, but an opportunity for free expression. We can engage without overthinking or feeling anxious about whether we did something correctly.

Even our figures of speech suggest this possibility. When we want to express that someone is capable of solving a problem easily, without the slightest effort, we say "it's child's play." When we call someone "playful," we imply an aura of happiness, that this person is able to see the world for its more positive, even more beautiful side. Humans are a lucky species to have the luxury to be playful. Some other animals also like to play, but I am certain that in each of us there resides a *Homo ludens* and that if the playful inner world of a person is accidentally dormant, it can be awakened sooner or later. At one stage of their life or another, everyone plays: the painter with their colors, the poet with words, and all the rest of us in the theater of life.

And of course, some people like to play with the Cube.

AT AROUND THE AGE OF THREE, a child begins to ask questions, and they nearly always begin with a "why?" Why are apples red and the sky is blue? Why are we not able to fly?

Why do we die? A child does not have be reminded about the Confucius teaching: "The man who asks a question is a fool for a minute, the man who does not ask is a fool for life." A child naturally lives it. We grow up, we learn how to answer questions, but in the meantime, almost imperceptibly, *we lose our ability to ask them.* Then, as we grow older and become curious in a different way, our world becomes more defined by the "hows." In a way, it is much easier to find answers to a "how" than to a "why"—perhaps because the "how" category almost contains its own solutions, whereas the "why" questions don't.

Questions define us as a species and also as individuals. "What?" and "Where?" are queries we share with most of the animal kingdom. Either prey or predator—these are questions of life and death. At the same time, only the apes, our closest relatives, and very few other species can join us in exploring "How?"—which leads to making tools for solving otherwise insurmountable problems.

Our everyday curiosity is rooted in such "How?" or "What if?" questions. This very spirit of curiosity can revisit old wisdoms and inspire questions about anything that previously had been taken for granted. The Greek philosopher Democritus's definition of an atom lived on happily for two and a half millennia. But then the question popped: What if atoms were, in fact, not the smallest particles and could be split further? How would one go about it?

In the same vein, solids had always been thought of as keeping their shape and breaking if twisted. But how to create an object that is a regular Platonic solid but is still able to twist and turn without falling apart?

It is the "How?" question that has defined most of my adult life, and it remains so, to some degree, even today.

How to write a book that doesn't resemble a book, for example? Or better: How to write a book while not writing it?

And finally there are the "Why?" questions, where humans are on their own (or so we believe today). "Why?" is always an abstraction, a theory that needs testing. It can be about others' intentions. It can be about laws of nature that need to be discovered. It can even be about introspection, trying to understand our own actions and desires ("Why bother writing a book?" is a difficult one to answer).

As I am getting older, I am spending more time with some of the eternal "Why?" questions of existence and mortality.

I have found that sophisticated knowledge is heavy—so heavy, in fact, that it can put brakes on the creative process. The more you know, the more difficult it becomes to remain curious. We have all had the experience of being in a situation where a so-called expert encounters an intelligent neophyte. It is easy to defer to the knowledgeable figure—so much confidence and learning is there—and not to appreciate the unexpectedly insightful questions from the nonexpert. And yet, questions from the amateur are often very original, becoming catalysts for new, imaginative solutions. In nearly every aspect of our lives, the most important and difficult task might actually be to find the right questions.

THERE ARE TWO WAYS to create change: Either find a new answer for an old question or find a new question that has never been asked before. It's hard to say which is more difficult. But clearly, the art of asking questions is one of the most important skills in our lives, and yet we never learn it in school.

I remember reading a Douglas Adams novel, *Dirk Gently's Holistic Detective Agency*, where Dirk says, "[D]on't you understand that we need to be childish in order to understand? Only a child sees things with perfect clarity, because it hasn't developed all those filters which prevent us from seeing things that we don't expect to see." I could not have come up with a better way of expressing what I have believed for my entire life. We all need to become more childish in order to understand more. As we grow older, the filters seem like thick vines and ivy that often cover up beautiful old buildings. Clearing away all the underbrush is a special challenge.

Learning is not about merely collecting knowledge; it's a wholly different process. In a way, knowledge partly consists of data, while learning is a skill that you achieve after you practice it over and over again. Soon you are able to do whatever it is more quickly and more competently. When one learns something, one collects both data and the skills to master that data, and the final product is knowledge. Knowledge is something deeper, not just the facts but their relationships, their connections to each other. Knowing how to handle whatever knowledge we accumulate through learning is very important. In fact, it's yet another layer added to the process of learning. In a way, it's like searching for something on the internet: You are capable of being online, but to find what you are looking for requires some skill, some ability to separate the useful information from the dross. Knowledge is how we can reach our goals, through a series of successes and failures. If we are lucky, we remember both.

Learning is a lifelong process, but it is most intense when we are children. How wonderful it would be if our ways of teaching were more sensitive to the best way of learning, which, once again, is playing! There is the old cartoon of all the students sitting in class and the teacher is pouring knowledge onto each student's head, a perfect description of a process that is neither teaching nor learning.

Huizinga noted that the history of the word "school" began with the Greeks, and at first the word meant a space for leisure and idleness, but "it has now acquired precisely the opposite sense of systematic work and training, as civilization restricted the free disposal of the young man's time more and more, and herded larger and larger classes of the young to a daily life of severe application from childhood onwards." That was certainly my experience.

I often thought that if I had had a different kind of education, I would be more capable than I am today. What do I mean by "more"? Am I trying to quantify something that is not quantifiable? I am not speaking of conventional measures of success. What I mean is that I would know more. My breadth of knowledge would be greater. Maybe I would be able to communicate in different ways. Maybe I would not hate to write because I would have been encouraged to write about many different subjects in many different ways.

My school was not able to capture my attention. But it did provide me with plenty of time to draw during class, and I consider that self-education to have been very valuable. But mostly, I was just bored. I retained what was personally interesting and forgot everything else. At that time in Hungary, students were required to attend a very high number of compulsory hours in school, six days a week, eight to ten hours every day. During my last year of elementary school, I convinced my indulgent mother that I needed to stay home on so many occasions that I failed to meet the required number of instructional hours. That meant, in order to move to the next grade level, I had to take a series of tests on the material.

It was the only time I earned perfect grades.

How can we encourage children to combine self-education with formal education? When we finish school, we typically don't know who we are. We don't really know what we know or what we are interested in or what we are capable of. Nor do we walk away with an understanding of how colorful the world is. Perhaps real education should hold a mirror up to children so that they can truly see themselves.

THE ASTRONOMER CARL SAGAN called *understanding* "a kind of ecstasy," and I think anyone who has had the experience of finally understanding something that once seemed difficult can relate to that description. To build a reality from seemingly disconnected data points, to see the consequences and gain the understanding, to solve a problem—these are other examples of what I call knowledge.

Often, the capacity to problem-solve has nothing to do with conventional measures of intelligence, as I have observed many times with those who are adept at solving the Cube. I have long hated the whole idea of quantifying intelligence with something like an IQ test and sympathize with those who think this test measures only the capacity to perform well on an IQ test. But this does not begin to consider some of the mysteries of real intelligence, which is fundamentally the capacity for making connections.

How could that possibly be measured on a standardized test? Obviously, it can't. But imagination is what will lead to the creative solution of problems. I still have a newspaper clipping from the Cube craze in the early 1980s. A letter appeared in the *Daily Express* in London from a mother about her fourteen-year-old severely disabled daughter who had learned to solve the Cube. "It's the first thing she's ever been able to do," she wrote, "that so many normal children can't."

I have never forgotten that, because it illustrates so beautifully what we may never fully understand about how intelligence might be expressed in unexpected ways. These are all intriguing clues, which offer no final answers, in the path to unraveling the mysteries of the mind.

THE SOMETIMES DUBIOUS MEASUREMENT OF IQ still has a unique appeal as the solid measure of intelligence, and often we forget about the other crucial aspect of human intelligence: emotions. If you wish, you can follow the current trend and talk about EQ, the emotional quotient, which is the capacity to understand human behavior, to feel, to decode the signs of the people next to you, not just by decoding the meaning of the words they utter, or understanding the concepts of any rational approach, but a much more complex and nuanced analytic talent. It is not enough to tell someone why it is important to do something. You need to be able to wake up some emotional resonance in them.

An object, its shape, its substance, its structure always has an emotional content. When you touch something, its tactile character has an emotional side. That it is made (or not made) of iron, wood, or paper. That it is not sharp, or it cuts your hand, or it inspires in you a friendly, familiar feeling, not to mention that perhaps it gives warmth, or it has a certain cozy sound, and that it creates, emanates a very distinct atmosphere. Of course, the Cube has an intellectual side as well, but his emotional appeal is essential.

High achievement in school does not always overlap with great achievement in life. We have all known brilliant people who did not perform well in school at all. I mean people like Einstein, who, legend has it, was not good in school and didn't want to be there. Though what he really disliked was the rote learning upon which his teachers insisted. There are those who are driven to be high achievers in school, but when they finally graduate, they do not have a clue about how to succeed in the real world. Yes, entrance examinations for higher education can be extremely demanding, but who is really able to judge an eighteen-year-old and evaluate what they will be capable of as an adult? What we can really measure is the effort it takes for kids to develop their skills.

A good teacher is beloved, sometimes feared, sometimes respected by their students. That emotional relationship in itself is already learning. A clever but boring teacher can't really transmit knowledge. We need the emotional surface, the very physical waves coming and going between those who teach and those who learn. First of all, it is an attitude that we have acquired in school.

And this gets us back to the question of solving problems. For me, the capacity to solve problems—creatively, confidently, and effectively—is the one important sign of true intelligence. Our relationship to those problems is also very crucial. They can be treated as a source of unhappiness that must be somehow addressed. Or they can be approached with the attitude described by the philosopher Karl Popper in *Realism and the Aim of Science*:

> [T]here is only one way to science—or to philosophy, for that matter: to meet a problem, to see its beauty and fall in love with it; to get married to it and to live with it happily, till death do ye part—unless you should meet another and even more fascinating problem or unless, indeed, you should obtain a solution. But even if you do obtain a solution, you may then discover, to your delight, the existence of a whole family of enchanting, though perhaps difficult, problem children for whose welfare you may work, with a purpose, to the end of your days.

When I read this, I felt that he expressed something that I had always known.

In the past, our formal and informal educations were the gateways that determined a profession we would have for our entire lives. All that has changed. Often the outcome is a matter of luck, of what we happen to discover, and which teachers take an interest in us. Sometimes, though, when we get a bit older, it is as if a subject finds us. We may wander without direction, surrounded by so many who appear to be like arrows rocketing toward a clear target, while we are more aimless, pursuing many interests but none of them turn into a passion. Until something happens: We meet someone who inspires us, or we study something that unexpectedly captures our curiosity. And then, we are hooked.

When I look at my grandchildren, I see that they have many more options than I had when I was in school. I see their parents looking at a variety of different types of schools and trying to decide if they should pick one that specializes in one-size-fits-all or personalized learning, digital or hands-on, standardized exams or creating inventions and products as a measure of learning, playful learning or theoretical learning.

You can imagine which ones appeal to me.

2

Imagination is the beginning of creation.
You imagine what you desire, you will what you
imagine and at last you create what you will.

—GEORGE BERNARD SHAW

WHEN I LOOK AT MY PARENTS, I see some of the same qualities and characteristics that I see in myself, but in different versions. I have my father's blue eyes and physical build, for example, but I have my mother's capacity for happiness. I think this capacity is something that one is born with—a question more of character than of circumstances. One person can be unhappy in prosperity, another happy in privation; one can be dissatisfied while successful, another does not lose heart even after a series of disasters. My mother was one of those people with a rare talent for a kind of optimism.

By most standards, her life was a difficult one, involving war, the too-early deaths of those whom she loved, physical and emotional pain, financial worries, and seemingly insurmountable obstacles to her doing what she would have wanted to do, which would have been to be a poet, perhaps, or a doctor. And yet, I never knew anyone more capable of joy at the slightest good thing—a delicious pastry, a good joke, the sunny sky. She had the gift of being able to easily transform tears into smiles and laughter.

She was brought up in the provincial town of Esztergom, in comfortable circumstances. She spoke fluent German and French and, as a gifted amateur pianist, loved music. Her dream was to become a physician, but that was not easy at the time, especially for a young woman. She was very petite, with a delicate face, a fine complexion, and a tendency for plumpness. Her best features were her brown eyes, which were shot through with green lights. She was an impulsive woman, one who would sometimes reappear shortly after she had left the room because she had just remembered something she still wanted to say or, halfway to her destination, decided that she preferred to remain where she had been. In some ways, she had too many talents and interests, including literary talents. As a young woman, she published a few slim volumes of her poems. (Unlike me, she did not hate to write.)

Everyone knew each other in Esztergom, especially if they were of roughly the same age and class. When my parents first met, my father had recently arrived there—young and handsome, full of energy and the promise of great things. My mother was a beautiful young lady whose first, brief marriage ended in divorce. She was quite sensible, but also very cheerful and mundane after her first marriage. Maybe their attraction was inevitable. But knowing them after the war, after my older sister and I were born, and after their marriage ceased to be happy and they divorced, it can be difficult to imagine how they even got to know each other.

I WAS BORN AT TWO IN THE AFTERNOON on July 13, 1944, almost exactly a month after D-Day, which marked the beginning of the end for the Nazi regime and World War II. Until then, as an ally of Germany, Hungary experienced the war only from a distance—except for its soldiers and of course those Jewish men who were forced to go into labor battalions already in the early 1940s. But by the time of my birth, the real fighting between the Axis and the Allied powers had reached Hungary in earnest. The Germans marched into Hungary on March 19, 1944, and about four months later the aerial bombardment of Budapest had begun.

My parents were living in two different cities, and my mother was caring for my sister while pregnant with me. She no longer had a family—she had lost both her parents and her only brother during the war. When she went into labor, she managed to get my sister taken care of and went to the hospital, which had moved many of its operations into a cellar that doubled as an air-raid shelter. She said I might have been one of the few babies to survive among the ones who were born in Budapest during that awful time, and this was the first indication that I was born under a lucky star.

In many ways, my father, Ernő Rubik Sr., was the opposite of my mother: He never seemed to be satisfied, even though he achieved a lot. My mother had a very open personality, but my father was very closed. By the time he accomplished what he had set out to do, his work was somehow inadequate for him. He always wanted it to be better. He found a measure of contentment, in small things: sharpening a knife so that it cut like a razor blade, or cleaning out the garden so that it was so pristine, one could eat off the ground. He had an unusual capacity for work.

My father was born in Pöstyén, a small town now in Slovakia. He was able to achieve a great deal despite having been raised in difficult circumstances. His father was declared missing in the First World War, and his mother brought up their three sons on her own, working hard. Scholarships made it possible for my father to complete his studies, both in elementary school and at the University of Polytechnic and Economy in Budapest. A group of young people met right after the university. One was a carpenter, another had some financial expertise and knew how to keep the books, and then there was my father, who was a gifted engineer.

They applied for a loan in order to start a small factory, which was organized around my father's expertise in designing gliders to be manufactured. In those days, gliders were used both for sport and for training pilots. My father had a great talent for refining the design in a number of different and very marketable ways. Their small business was a success and they repaid their loan right before the war.

Then all was gone.

After the war, the factory was nationalized and my father remained there as an employee. This was a very unusual arrangement. This was a time when private properties, houses, apartments, land, and factories were seized by the new, Communist government, but my father's professional accomplishments had made him indispensable. He became chief engineer of the factory he had founded. His expertise meant he was also the engineer, the designer, and the marketer; he also named products and even operated what the factory produced. This wasn't normal in those days, when former factory owners, who did not escape abroad, were jailed or demoted to menial jobs.

There are many questions I wish I had asked him about his work and his life. But he was not a man who told stories. The past was not a time that was pleasant for him to remember. And so, I never asked. We were so independent, we shared very little except our nature.

In the early 1960s, he created his most famous aircraft, the R-26 Góbé, which became the classic training glider in Hungary (it was exported to Cuba, Austria, and the United Kingdom). He was awarded international prizes from aeronautic organizations, and the Kossuth Prize, at the time the highest honor bestowed by the Hungarian government. His gliders were made of aluminum, so they were very light and agile. Flying was a popular amateur sport in Hungary, and most people who had any connection to flying knew his name. After he died, a street and an airport in Esztergom were named after him. Sometimes, even after his death in 1997, when I walked my dogs in a park near Hidegkút, a suburb in Budapest near a small airport, I could still see his gliders flying in the sky.

I have a strong impression of my father's hands. They were the hands of a worker, strong, large, with the nails cut very short. They looked somehow incongruous holding a pencil, as if he could easily snap it in two with a casual move. Even when he was officially retired, he never stopped working; he sat surrounded by plans and designs for less expensive, even safer small planes. I would like to be able to say that he took me flying, but he never did. My mother would not allow it.

In their early years my parents must have been at least a perfect example of the old truism "Opposites attract." My mother was a wonderful talker with a resonant and perpetually youthful voice, which she was skilled at deploying. In nearly any situation, with family, friends, or even a stranger, my mother would plunge into conversation with enormous enthusiasm, making the most of it and never letting it languish. She loved the richness and flexibility of the Hungarian language, and it was a great blessing for me to have acquired my mother tongue from someone with such a sophisticated understanding of its nuances and possibilities.

My father's response to my mother's chatter was mostly silence, with some monosyllabic responses when absolutely necessary. His language was practical, economical, concise, object-oriented, the language of the workshop or factory floor. There were no flights of fancy with my father, no ruminations about possibilities. He restricted his observations to concrete facts and realities. Plans would be realized through hard work, not pointless discussions. Emotions were best left unexpressed.

If I learned to appreciate conversation from my mother, I learned the art of silence from my father. Not as a passive act, but instead that there was something important about resisting the urge to fill up a space with conversation just to avoid silence. Like an artist who experiments with negative space in a painting, I have found that within peace and quiet, much can be accomplished.

I also share my father's perfectionism on some level. When I find that my efforts fall short of my goal, it only pushes me to keep going on. In contrast to my father's more pervasive dissatisfaction, I am able, at least privately, to savor a sense of accomplishment. Or perhaps I am just more selective in what I do. His last project was to design an ultralight plane that used only human power to fly. It was very late in his life and he never finished it. My earthbound goals may be more accessible than his desire for the most perfect flying machine.

FROM THE TIME I WAS SMALL, I loved to draw and paint. I liked to use paper, pencils, crayons, watercolors, and oil paints to try to capture what I saw in the world around me. I liked putting it on paper, and enjoyed the results. For secondary school, I attended art school and concentrated on sculpture. It wasn't because art school was part of my family tradition—though I discovered much later that one of my grandfathers was an artist—just that it seemed the natural extension of elementary school.

Those four years were enough to convince me that I was not a "real" artist. I felt somewhat out of place among my classmates, most of whom were rebels and bohemians. And I felt restless in a way that being a sculptor did not fully satisfy. I wanted to combine my artistic interests with something practical.

After graduation, most of my classmates went on to the Academy of Fine Arts, but if I was sure of anything, it was that I did not want to do that. So, I enrolled at the Budapest University of Technology to study architecture. I am not sure how I chose architecture. My father being an aviation engineer probably had something to do with it. Maybe I was also drawn to it because geometry came so easily to me. I enjoyed studying architecture, and it was a department that found its students often working late into the night on projects that felt quite demanding. And yet, when I finished college, I felt completely unprepared to work as a practicing architect. When in doubt, there is nothing wrong with a little more education, so I applied and was admitted to the Hungarian College of Applied

Arts, which at the time occupied a special status, somewhere between a university and an art school.

The school was small, with only about a few hundred students in total, and because of its size, students and teachers really got to know each other. We worked closely together, not so much in classrooms but in studios. The textile, ceramics, and metal workshops were all available to us. I had found my profession, my place. And naturally, once one finds where one is supposed to be, one performs well. It was during this time that my learning was transformed into teaching. Even before I received my degree, I was offered an assistant lectureship.

And almost without having noticed it, I gained a profession. I taught architecture and design and also was involved in working on buildings. I had never dreamed that I would become a lecturer. But after I got the offer, I thought, *Why not?*

I enjoyed this environment. As a teacher, I was allowed to continue my education in the studios. I liked the school and did not take life very seriously. I was not a young man who was driven by some burning ambition to make some kind of mark on the world. But I was supporting myself, and even without being consciously aware of the situation, I was fortunate enough to have a job that did not feel like work.

During my studies, I became very interested in the different kinds of patterns I could create with geometrical shapes in my drawings, paintings, and sculpture. As a final project at the end of art school, we collected some of our work from our three years of studying and installed an exhibit. When I selected my work for such an exhibition before my graduation in 1970, one of the pieces that I chose to display was a colored cube.

EVEN THOUGH I HAD A PROFESSION, I was, and still am, an amateur. Amateurs are often considered to be distinct from professionals by their education. If most professionals are educated in their field, then most amateurs are self-trained in the area that captures their imaginations and interest. But paradoxically, those who are at the top of their professions are very close to amateurs in their unconditional dedication to their work.

It is fitting that the etymology of the word "amateur" comes from the Latin word for "lover," which is *amatore*. No matter how far away the word may have traveled from its root, it still suggests to me that you love whatever it is that you are doing. You love your subject. You love the process. You love the result. In contrast, the work of a professional is quite different—and it usually involves financial compensation. Naturally, I respect professionals; I just don't consider myself to be one. Professionals must be detached, purpose-driven, must use well-proven means to well-identified goals. On the other hand, amateurs enjoy an almost absolute freedom that a professional never will. An amateur is free, independent, and open, with all the downsides of making accidental discoveries, but that is the reward also. Amateurs tend to be immersed in a state of curiosity and discovery. They enjoy the process, the small revelations of progress, and the flash of genuine pride in the final result.

Looking back to the past, when the story of the Cube began, my experience suggests that you can succeed as an amateur. When I made the Cube, I was not an industrial designer, and I had no expertise in the toy area. And I worked totally alone. I was the amateur inventor, I was the amateur engineer who had to solve the technical part of the idea, and I was the amateur designer who finalized its shape and form, its appearance.

When I first attended toy shows, I noticed that I had something important in common with nearly all the other "toy inventors": We all had different day jobs, doing something else to make a living.

I discovered that one interesting characteristic of most of those who have succeeded in the toys and game field is that they tend to be amateurs and autodidacts.

Then again, even in my attempts in these sometimes clumsy descriptions, I am confronted with the limitations of my writing. Put on paper, a text can become so dry and constrained, the real flavor, the nuances, and the emotional content of real life are nowhere to be seen. It is as if you have a collection of dried leaves, carefully pressed between the pages of an unread (or unreadable) book. The leaves may be beautiful examples of nature's remarkable variety of forms and shapes, but the richness of the different smells and the original colors? Gone.

You see, these two terms—"amateur" and "professional"—may seem to be contradictory, but they can exist simultaneously in us. Maybe the best we can hope for is to be both at once: a forever amateur professional, or someone who makes being an amateur his profession. The key seems to be to enjoy our professional activities, so as not to lose the way we felt when we enjoyed our first achievement, and to anticipate every new task with the enthusiasm and zest of the amateur.

But this is not the case for me. To be honest, fortunately (or unfortunately), I don't think I have become a professional in anything. I am and remain an amateur in everything that I undertake.

CLOSE YOUR EYES AND IMAGINE an object in as much detail as possible. It can be anything, but my advice is to pick something as simple as possible: a table, a chair, a coffee cup, a vase. These are concepts that exist both in a physical reality, as well as a simplified image in your mind. And now, once you have this visualization, once you really can hold this picture in your mind, try to make it move. Take a stroll around your object. Or maybe see if you can rotate it in front of you. Become an ant and crawl over it and under it, examining its every angle and surface. Turn into a hummingbird and fly around it and fly over it. Finally, peek inside the object. Can you see the structure that holds it all together?

Of course, you may have found this little exercise difficult, even frustrating or, worse still, pointless.

But do not be concerned; you are in the majority.

The inner sight that this exercise requires is neglected or mostly ignored by our society and our system of education. We live in a tactile and experiential world, so what is in front of us is typically noticed but not internalized. Understanding that, let's move a bit further in this exercise. Now, try to make a sketch of what you had seen in your mind. You don't need to make something precise, or beautiful, but make something recognizable. Drawing needs to provide a way of giving meaning to something, less than it requires what is considered to be "a good eye."

ONE OF THE CLASSES that I taught when I started to work at the Academy was descriptive geometry. In one way, this could be considered a dry subject, but it is very connected to visual communication, design, and art. It is a subject where students have very little existing knowledge, so it can be difficult to teach and to learn because you need to master the principles and laws with a specialized vocabulary. At its most basic, descriptive geometry involves generating two-dimensional views and descriptions of three-dimensional objects. Explaining these concepts was never easy, but it prepared me for my work with the Cube.

The Cube has a different kind of spatial orientation than most objects—there is no up or down, right or left. One reason for its enduring appeal, for its capacity to engage all sorts of people in all sorts of ways, may be that it is something that is a whole, but it is impossible to truly see all of it at the same time. (Maybe it's a bit like another person—you never are able to fully see someone else in all their dimensions.) No matter which way you turn it, you will only have a limited perspective. The challenge is, you need to be able to see all the sides in order to know if you are solving it. One side may be a single color, and you can bask in a moment of accomplishment—until you turn the Cube and see the chaos on the other five sides. All the time, your view gives you just one still-life perspective, but you need several perspectives to put together its position in space.

Here is where your imagination and memory come in handy, because even though you can't see it all, over time you can train your mind to remember the useable part. When we are children and our mothers show us the broken pieces of the vase that fell when we tried to climb onto the table, she reminds us that our actions have consequences. As we grow up, we start being able to predict those consequences. So, too, with the Cube. After we play with it for a while, we start being able to recognize what will happen when we make a turn.

MY CHILDHOOD BOREDOM at school prepared me for my interest in geometry. During the many days that I spent drawing my surroundings and objects as various teachers droned on, I started to get a much better understanding of those objects' essential nature. I began to realize that for me, drawing was a way of understanding things. Behind the picture there was content, and when drawing the image of something—a face or a tree—I could try to capture its appearance. But without understanding the nature of the structure beneath that surface, the image remained flat on the page. This was the kind of insight that can only arise after feeling some frustration when you are clumsy and unskillful in doing something that is inherently interesting for you. As you begin to see the flatness on the page as a problem to be solved, there inevitably comes a moment when the solution makes itself evident.

Descriptive geometry has its own language, a language both powerful and simple. Just as writers usually do not invent words, I did not create this language. What I did was discover new possibilities within the spatial relationships that form its vocabulary. Descriptive geometry is about understanding space and using that understanding to both discover 3D ideas and become capable of communicating those concepts.

I know that as I became more and more involved in the arts, I suddenly realized that in order to draw a person, even without being able to see their skeleton, I needed to understand the anatomy of the human body. An inanimate object is no different. It, too, has an underlying structure and without it, whatever is drawn is just lines on a page. This is where descriptive geometry comes in, because at its most basic, it involves developing the habit, or the skill, of looking beneath the surface at the true nature of things.

YES, WE ALL MAY BE BORN WITH THE CURIOSITY, even the talent, to look below the surface of things. But it tends to diminish over time, because there is an epidemic of space blindness, and other skills are made to seem more important. Especially since the level of understanding—to grasp, penetrate, fathom, recognize—the underlying structure of a human being or an object is something that cannot be quantified. How do you grasp, how do you fathom, how do you recognize something? It can't really be measured or tested in the same way that someone's knowledge of names, dates, facts, and figures can. When I considered the underlying structure of what I was drawing, my awareness and understanding not only of an essential nature, but also of connections that existed among the various components of that object or person, began to develop.

I mentioned space blindness, but I should be clear about what it is. It is the inability to understand spatial relations and the lack of the capability to become oriented in space. Any kind of specialized knowledge creates its own language—just consider how difficult it often is to understand what doctors are saying. Every area of specialization is accompanied by a new vocabulary that is often completely incomprehensible to an outsider. But that new vocabulary serves a very important purpose: It becomes a simple way of expressing a very complex concept. Without it, many words would be needed to describe what one is talking about.

If I didn't train my mind in descriptive geometry, I would have needed many more words to describe a single 3D relationship, when the easiest way to describe it would be by making sketches. Imagining only one line in space can be very difficult. And if you have more than one line, it becomes more complicated. If they are crossing each other, they are easy to imagine. If they are parallel, it is also easy, since we have so many analogies. But lines that cross share one point and close an angle. Parallel lines have not only one but two shared points in both directions to infinity. Both cases—crossing lines and parallel lines—define a plane, which means they are

2D. Then there are lines that have no shared point, which can only be described in space. The lines themselves are not three-dimensional, but in this case they define 3D. If you would like to describe the relationship of these lines, you can only use the terms of 3D. In 3D there are many things that don't exist in the 2D world, volume for example. But it is important to note that the distance of these lines can be specified only in 3D by a third line crossing both at right angles. When I taught, I found that this would be a simple but efficient way for a student to check their knowledge and vision of 3D. If we go back to spatial imagination, I can create tasks to try to change my students' ways of thinking about their surroundings.

To have fun is important. Year after year, I tried to evoke a feeling for this in my students. I tried to teach this subject like a foreign language, vocabulary and style hand in hand, to bring home that the way we speak depends on what we have to say, and we must choose our words and tone and rhythms accordingly. When we have mastered the language of descriptive geometry, we can communicate something different, with more accuracy, about the space that surrounds us and about the objects that define this space. We can only comprehend the structure of space by understanding our limits and possibilities.

I often felt as if I were talking about light and color to the blind. This isn't a subject that can be learned or taught easily. Indeed, space blindness is very difficult to cure, especially in adults who have spent their entire lives establishing their own very definite pictures of the world. Creative work is an expression of our most intimate lives, what we feel, what we believe, the way we want to proceed, the goal we would like to reach. In the same way, trying to understand the most abstract notions of space and geometry involved very precise language and communication but also a highly trained sensitivity to concepts that, well, either make sense or don't.

HOW, THEN, DOES SOMETHING NEW, something that up until then had not existed, become a reality? What does creation consist of? Indeed, what does "to create" even mean? These questions have answers that are the catalysts for even more questions.

But we must ask them anyway. Of all of them, perhaps the questions that I'm truly qualified to answer are these two: When I was working on the Cube, what did my creative act look like? What was my experience of it? I can't say that what I did was the result of a lightning bolt of inspiration—another illusive eureka! moment—where almost out of nowhere I could imagine such an object and all else followed. There was no definite point in time that I can identify as a moment comparable to Newton's proverbial experience with the apple. No chance event or dream or encounter prompted it.

I was just tremendously curious. I wanted to find something out. Even if I didn't know exactly what that "something" was.

I began to work on a certain geometric problem, and my goal initially was—nothing. I just had an idea but was not even able to formulate precisely what the problem was that I was attempting so ardently to solve. And yet, the potential of coming up with a three-dimensional object in the shape of a cube that could move on its axis became appealing. Because of that, I started thinking of all the next steps. Soon, I kept on playing, as I would with any interesting game. If something is interesting, it wakes you up. There can be a competitive aspect, the hope that in some way the game can be won, and that becomes a driving force.

I had no deadline. I simply enjoyed the problem I had stumbled upon and knew that, for me, working on a problem is a prerequisite for fostering—or liberating—my imagination. Eventually, of course, I found answers to these questions (or better: the answers found me) in a 3x3x3 item with red, white, orange, green, blue, and yellow sides. And that's it.

Curiosity means that we accept nothing, that again and again we question the fundamentals. We find the good questions because we are interested in the "how" of things. That is the only way forward. And here "forward" doesn't mean a predetermined direction or something that can be fixed beforehand, but the openness, not just of the eye, but of the broader vision. To introduce another paradox: Moving forward requires a kind of sophisticated and well-prepared naïveté, or the unquenchable thirst for ever better answers.

But not simply theoretical answers. Creation needs something more than curiosity; it also requires a kind of inner drive or ambition that can be characterized as curiosity about one's own capacities. Do we know what we are capable of—really? Like the ancient man, one works with the elements or stone, wood, fire, using a hammer, nails, a screwdriver, and all the instruments that humanity has since developed. But the other key ingredient is intuition. The simplest instrument itself can, through its use, inspire a solution.

Curiosity is the flame that can ignite creativity.

3

When I am working on a problem,
I never think about beauty. But when I have finished,
if the solution is not beautiful, I know it is wrong.

—R. BUCKMINSTER FULLER

IT WAS THE SPRING BEFORE my thirtieth birthday, in 1974, and my room was like a child's pocket, full of marbles and treasures: bits of paper with scrawling notes and images, pencils, crayons, string, small sticks, glue, pins, springs, screws, rulers. These items occupied every corner, the shelves, the floor, the table, which doubled as a drawing board; they were hanging from the ceiling, pinned on the door, tucked in the window frame. Among them were innumerable cubes: cubes made of paper, of wood, in monochrome or colored, solid or broken down into blocks. Ostensibly, this was where I prepared for the classes I taught, and where I could come up with some ideas for my students. But it turned out to be much more.

One day—I don't know exactly when, I don't know exactly why—an idea took hold of me: I thought it would be interesting to put together eight small cubes in such a manner that they remained joined together but were also capable of being moved individually. I hadn't the faintest idea whether it would be interesting to anyone else. I was preoccupied by the seemingly unimportant fact that it appeared to be absolutely impossible to give all eight pieces free and independent movement while they were still connected.

A mechanical relationship or correlation involves a measure of constancy. A door always turns on its hinges: it moves, but never moves away from those joints. The wheels of a car always turn around their carriage axle. The first question that faced me was figuring out how to keep the eight small cubes together so that they could be both connected and also move while they remained connected. Theoretically, four of them could turn around the rest at the same time, but where should its real axle be? It was quite easy to make a model to illustrate the problem, so I took a few steps in doing this.

First, I made eight identical cubes out of wood. Their edges might have been filed smooth—I don't think that initially mine were, but it's nicer to picture them that way. Then I drilled a hole in the corner of each cube so that I could link two cubes together and then obtain four pairs. Once I had four pairs of cubies, I then linked the pairs at the opposite corners so that I created a little block, a 2x2x2, made up of smaller blocks, whose faces I could turn both independently and when they were connected. I felt as if my problem was solved! I had made something marvelously simple that appeared to do what I had intended it to do, which was to have it be self-contained but capable of movement.

In short order, it became clear that my masterful construc-

ERNŐ RUBIK

tion was no solution at all. It literally fell apart. In the center I found a big and unbelievably tangled knot. The elastic bore the strain for some time, but the turns proved to be too much for it and it broke. It was frustrating, but I was very curious about *why* this happened. I had given shape to the nature of a problem, and this three-dimensional problem had gotten hold of me. I was hooked. Both intellectually and emotionally. The problem was not just a theoretical question: It became a physical presence in my hands.

I was not dealing with an abstraction, I was not dealing with a concept, although it did start out like one, and in my mind still did feel like one. This was something real. An object like a concept and a concept like an object. Somewhere in between the two. A sphinx of a problem. If I have the key to the object, I shall have the key to the concept as well. I had to overcome an unknown distance in a yet unknown land. How embarrassing. How inspiring.

Over the next few days, I began to investigate the nature of the problem and its possible solutions. This investigation phase was carried out in my head. I discarded various possibilities on theoretical grounds without even trying to execute them; they seemed too clumsy and too complicated. I was convinced that there was a solution, and that the solution had to be simple.

Right now we are speaking about structure, or more precisely, the construction of an object. In a construction sense, I created a connection with the cubes, making the corners all hold together—until they fell apart. I realized the possibilities of movement were so rich, the connection with the rubber band by itself was not workable for long. It could maintain its structural integrity for several turns, but not hundreds. I needed something that was more durable, that had almost endless capability. There were two essentials: It had to have the axis and the connecting elements. I replaced the rubber bands with fishing line, but it was still not the ultimate solution. It required rigid elements, and I then attempted to make it slightly more complex, but also simpler. I separated the two functions: fixing the rotation around one axis and creating a cohesive force for the pieces.

I CONCLUDED that the solution would be provided not by 2x2x2, but by 3x3x3. It was clear that already the technical solution needed more pieces: 3x3x3 contains 2x2x2, but besides the corners, it has also the middle parts and the edges. With middle parts and a hidden center, the movement can be progressively more complex. It becomes an answer to the original question, which was a query about how a structure could be created in which the singular elements are joined together but were still capable of being moved individually. This is how my first wooden model was made—and it worked! The model, however, had only twenty-six pieces. At first I didn't think that the center cube was necessary as far as the construction was concerned.

I eventually realized that it was essential; *indeed, it was the very core that would hold it all together.*

What I had made was clearly an object, but more interestingly, *it was a three-dimensional embodiment of a concept.* As a representation, it only contained the *essence* of a three-dimensional construction, just as a pictorial representation contains only the essence of the real image. Constructing models is an essential part of a designer's activity, whether as an architect, an object-designer, or someone working in any related field. But models made in the course of that kind of work served only as illustrations.

There is an ostensibly minuscule but in fact infinite difference between the reality of an object as it exists in the world—be it a building or a tennis ball or a cube—and its geometrical perfection. Everything in the real world is flawed just a little bit in comparison to its ideal geometrical counterpart. This is a momentous, even an enormous detail. Geometrical definitions are crystal clear, but the physical expressions of those definitions don't exist in the real world. Even the smoothest mirror will never, ever measure up to what we mean when we say "plane."

I had made "perfectly" regular cubies, and yet they still had unavoidable, minuscule variations. Nothing anyone could see. But in comparison to the ideal that exists on paper—or now on the screen, or as a definition or theory—they fell short.

I MADE THINGS EVEN MORE COMPLICATED by wanting to have *movement* in my structure. Movement naturally changes the relative position of an object, while regularity remains a force that maintains the object's integrity and stability. I wondered if it would be possible to include both regularity and movement within one structure without, say, the movement wearing away the corners. This fluidity can only be achieved to a certain level, a concept known as "tolerance" in technical terms. Tolerance is a plus or a minus that is used to acknowledge that whatever is represented is not precise, but the resulting structure is capable of *tolerating* the differences. We can come close, we can aspire, but complete accuracy will never be achieved. Maybe "accuracy" is not the exact word here: I was looking for an ideal object that could contain my vision of this contradictory function, a constant and fixed position combined with the capacity to change this position—and I got as far as the Cube.

All of these theoretical concerns eventually became practical when the time arrived to manufacture the Cube. But first I had to relish the moment when I looked at the problem that I had solved: at the small object that I had almost involuntarily created. I wondered about what *other* problems this object might present that needed a solution.

PROBLEMS ARE INEVITABLE. They are an integral part of life. As a general rule, they don't go away on their own. Sometimes they can make us feel quite mad, in both senses of the word: crazy and furious. Many times, they teach us important lessons. It has often been observed that finding the solution to a puzzle is a kind of microcosm, a model for problem-solving in other parts of our lives. When we are not disorganized by the challenge or made to feel too anxious by our sense that whatever we are facing cannot be solved, we need only break a problem into parts, solve each of those parts systematically, and then put them back together. Then we start to fully understand the nature of the problem, which helps us to solve it, and, more important, to be able to understand completely what we did in the first place.

This seems like the perfect opportunity to refer to my compatriot György Pólya's classic work, *How to Solve It*. Here is what he said.

First, you have to understand the problem. After understanding, make a plan. Carry out the plan. Look back on your work. How could it be better?

If this technique fails, he advises, "[T]hen there is an easier problem you can solve: find it." Or: "If you cannot solve the proposed problem, try to solve first some related problem. Could you imagine a more accessible related problem?"

And to these pieces of advice, I also can add: If you deconstruct an object, you have to be able to put it together again.

I LEARNED A GREAT DEAL ABOUT problem-solving and perseverance as a child, when I watched my father build our cottage on the shores of Lake Balaton with his own hands, from scratch. My parents bought a plot of land on this beautiful lake in the fifties and, at first, we slept under blankets in tents, drawing our water from a well, cooking over a little fire, and relying on a paraffin lamp after dark. Slowly, my father began to construct our cottage, stopping there whenever he could to work on it. First, walls appeared, and we lifted the blankets from the ground and used them for doors. Soon, actual doors appeared, then rudimentary plumbing and electricity. We did it all with our hands. It was because of financial necessity, but this process helped me get so much closer to the material foundations of things. A house for me is not simply a house one moves into when it is ready. A house is a human construction that needs infinite attention to all the details.

As an architect, I have come to understand that a home is, or should be, much more about the personality and character of its owner, or whoever plans to inhabit it, than of its designer. And this leads me to a question, which I explored in one of my essays about the subject: *What is beauty? Is beauty useful?*

Any new product, physical or virtual, is also necessarily a new design. However, it is far from self-evident that the resulting "phenotype"—the way it looks, its observable characteristics or traits—is the most perfect match of the "genotype," which is the code, its genetic makeup, the functional construction of the product itself. Phenotype is the expression of an organism's genetic code, or its genotype, and the influence of environmental factors, the mysterious alchemy of so-called "triggers" and chance. Such evolution takes place over several phases (or generations) of improvement, trial and error, and the feedback of disenchanted users. In the rare cases when the harmony of function and design miraculously come together, beauty is achieved.

This beauty is cathartic because the inherent contradiction of function and experience is resolved. This is what the late Steve Jobs so perfectly understood. But then why did it take Apple so many years to realize this (expensive) quest for beauty? Why is design becoming increasingly important for mass-produced products that are meant for mass consumption, as opposed to the exclusive few with elaborate tastes and thicker wallets? And why is it only in this new millennium that the importance of design is realized both in business and in education?

I think these questions are partly explained by the decreasing marginal utility of sheer performance. Our computers, as well as our cars and television sets, have become so powerful, that adding more gigabytes, larger Winchesters, more horsepower, or more pixels has become less and less significant for everyday users. *The competition for the consumers' attention and satisfaction has quickly shifted toward a richer experience where beauty is key.*

Another reason is that the information flow in our interconnected world makes this competition fiercer and nearly constant. The Web, furthermore, gives more weight to the acquired taste in any given category; regardless of the price tag, an object of beauty quickly becomes an object of popular desire.

At least in principle, functionality can always be improved—it is at the harmony of function and form where design can come close to perfection. A product, or even an object of art, is only perfect when there is nothing more to add and there is nothing to take away. This is the cathartic experience of an object becoming itself.

The Cube became iconic because of its counterfactual functionality: It made something possible that was seemingly impossible by cracking the inner immobility of a static solid. Just as important, however, it created a harmony of the mind, the heart, and the hands in dimensions that were fit for manipulation, a task that required cognition involving colors that evoked immediate emotions. It is also an object in and of itself because it sets its own challenge: a puzzle that needs no instruction manuals or elaborate rules. Anybody blessed with the basic human senses anywhere can instantly "get it."

So far, so good. But once the relevance of design is established, what is to be done about it? First of all, the word "design" is already problematic. It is painfully overused, which blurs its meaning. For designers themselves, it refers to creative challenges, user experience, functionality, and appearance. For them, it's something to work on, a present-tense verb.

Design has a rather different meaning for "laypersons." Rather than an action or an activity, it's used to describe something being cool, trendy, or appealing. For them, it's a self-contained noun.

The task is to close the gap between these two different meanings. In order to accomplish that, we need to realize the interdisciplinary character of design and design education. In this sense, design (as a creative activity) differs somewhat from other fields, where discipline implies a certain depth of understanding in a specific context. For example, listening to a symphony is a different experience for the ordinary concert-goer than for the music scholar or professional.

Design, in contrast, is interdisciplinary by nature. The end goal of the design project is not merely the object, but the object in use, and quality can only be measured by the interaction of the object and its user. This quality may only be achieved by the joint understanding of the human content the design appeals to—the psychology, the perception, sometimes the anatomy, and even the economics—and the character of the object, which includes the materials that are used, whatever information technology might be involved, its mechanics, and engineering.

Therefore, it is the coming together of human content, technology, science, art, and creativity that resides at the very core of design philosophy and should also be at the center of design education. Because of the overwhelming span of creative design perspectives, design education must start from the basics: hands-on exercises with real materials, a rigorous understanding of dimensions of space, the workings of 3D, and much more. Ideally, design education should begin as early as elementary school so that the human experience itself informs design professionals and makes them open and available to creative challenges in interdisciplinary contexts.

At its core, design is the link to nature for artificial objects. Nature does not know strict borders or barriers; it only knows transition. An understanding of the various contexts and connections and opportunities of transition is the very heart of inspiration and creativity. In order to vindicate its meaning and relevance, design must rise to this challenge.

I NEED ONLY LOOK BACK ON MY FATHER'S CREATION of our little cottage on Lake Balaton to see all those forces at work. It turned out to be a small space but adequate for our family, with a closed courtyard and a veranda, and a separate summer kitchen that originally was supposed to be only temporary; but then, as so often happens, temporary plans become permanent fixtures.

The result reflected many of my father's most admirable character traits. He had intended the cottage to be much bigger, but his plans would have taken an eternity to complete. My sister, mother, and I each made our own contributions. I would help my father mix the cement, water, and gravel, and we poured the concrete together. I learned how to recycle old nails by hammering them into shape without smashing my fingers. When the house was nearly complete, I made a floor mosaic for the veranda.

We never built the additions he had envisioned. The rusty iron girders that were evidence of his plan jutted out of the walls and gave the cottage an unfinished quality, as if these were the markers of an ongoing construction site.

As it happened, my student years were punctuated by the long summer vacations at Lake Balaton. Balaton, called by some "The Hungarian sea," by others simply "The Lake" was, in those days, still very romantic, its waters sweet and clear, its shores wild with reeds and game birds, its south shore flat, its north side hilly.

One summer my father even bought me a sailboat—a small secondhand craft—and from then on I spent day and night in it, in utter solitude. I made a small harbor for the boat by cutting a path in the reeds next to our plot. I loved to go far out to the middle of the lake where no noise reached me from the shores, listening to the occasional splash of a fish, gazing at the white sail and the white furrow cleaved in the lake by my boat, with the water cooling my feet and the sun burning my skin. And I adored the storms.

It is a gentle lake, so it catches one unaware when it decides to play rough. Its storms are sudden and fierce, and only the natives can read its warning signs. Its sly rages cost many outsiders their lives. The northern hills of Balaton delay the winds effectively, only to drop them on the water like a bombshell. The transformation takes place within minutes. There is darkening in the water, then a white line races towards you, followed closely by a pitch-black mass, which only a second ago was an innocent, gentle green. By the time you notice anything, it is upon you: a peculiarly shrieking wind, the air sharp with spray stolen from the turmoil. Even seafarers are amazed to see this apparently harmless, complacent creature erupt in such sharp, treacherous waves, crests frozen into peaks of fast heartbeats. The water remains warm, so it is still marvelous to swim in. The next best thing to swimming during a storm was doing it at night, with the mirror-smooth surface broken only by the long trail left by your own strokes. You are so reluctant to get out that, in the end, you find that you have "used up" all the water underneath you—still loath to leave, you crawl in the soft mud until it casts you out on the shore. When not on or in the water, I went on long hikes around the lake on my bike. It takes two days to make a full loop, but you can cheat and take a ferry across midway.

This childhood paradise by the shores of Balaton is long gone. In retrospect, I marvel at my father's physical strength and persistence to have built the cottage there. To me, that little house is also a symbol of my father's strong belief that problems—even those contrived by nature itself—only arise in order to entice us to figure out new and original solutions.

There is a huge difference between defining the problem and solving it. Most of the time, situations emerge from a kind of chaotic state, and most of the time it is counterproductive to be systematic about even beginning to solve these strong dilemmas. But our whole life is about problem-solving: One problem solved, and another pops up. No matter how disorganized they may appear to be, the first step is to find some small fixed point in the chaos, to get a foothold and create some almost imperceptibly small foundation of an order where we can begin to address the whole.

Thousands of books were written about how useful chaos can be for an artist's intuition, how disorder is inspiring. Or laziness as true inspiration. Yes, sometimes the opposite of it is also true. From my perspective, a well-founded system of life encompasses a catalog of things laid out clearly, but still, one that has the capacity to gaze fearlessly into the eyes of chaos and accept the fact that not all things make sense all of the time. If one has the ability to connect the distant points, chaos is the most inspiring challenge in the world.

AT LAST, I HAD MANAGED TO PUT all the pieces of the 3x3x3 wooden cube into place, and the construction appeared to be stable. Imagine the Cube as you know it, but monochromatic; at this point, all the sides looked the same. Bleak, abstract, flat. No vivid colors inform you of your actual position. I wanted to see what would happen if I moved it. I imagined moving the top layer not in a full revolution, but just halfway, by forty-five degrees. What would occur at the end of that move was clear, but I was also curious about that in-between state when the turn was not complete and the edge-cube had already left one arm but had not yet reached the next.

I could see that, in this position, the edges were connected to only one middle. I realized they were held also by the edges in the middle layer, which were still held by their two middle neighbors.

There was an interdependence among all the parts that I had not expected but discovered was an essential part of the structure of what I had made.

It's strange, isn't it? The Cube suggested things to me that I hadn't anticipated, even though I had created it.

I sat with this new object, and I started scrutinizing its hidden capabilities, both in terms of its individual components and also its collective identity.

I returned to my original, wooden model and realized why the core cube that at first I had omitted as unnecessary turned out to be indispensable. The center is the single intersection point for all the axes. There is an element there at the core, with screws and springs that give each of the individual pieces of cubes the power to push or pull. The center creates a tension analogous to gravity. In the same way that gravity keeps us earthbound, the springs are drawing the middle pieces toward the direction of the center, but in an elastic way. When I began, I didn't think there was a need for a functional middle because I was using the rubber bands to provide the tension that kept everything in their places.

The rubber bands, however, had two major problems. First, they were not a durable constructional element, so they basically didn't last very long before they fell apart. Second, the tension they created among the cubes was not strong enough. The whole structure became too malleable, like a rubber ball that wasn't inflated enough. Then I created a middle element in which I drilled holes on all the sides and attached screws to the holes. The trick then was to calibrate exactly the right tension so as to make it possible to turn the different elements with ease but not too much rigidity. It needed to be both stable and adjustable. When I turned the screws tighter, it created more tension; when I loosened them, they became once more like a ball that had lost its air. To make my square ball harder or softer was all a matter of turning the screws. (Years later, speedcubers would also adjust this to make the action of the turns what they desired.)

There is a famous story about Beaumarchais, the playwright and adventurer, who in his ripe old age created the immortal figure of Figaro for the world stage. When he was an ambitious young man of twenty-one in July 1753, this son of a watchmaker, who was endowed with the unquenchable drive to climb higher on the social ladder, spent a year inventing an important "escapement" for watches, which allowed them to be substantially more accurate and compact. At the time, pocket watches were commonly unreliable for timekeeping and were worn more as fashion accessories. The royal clockmaker, named Lepaute, first encouraged the young man in his efforts but later simply stole this very useful invention, thinking that his own reputation, his sole word, would have been sufficient to secure the renown for him. But the French Academy of Sciences, in a landmark decision that surprised the whole world, ruled an unknown young man was responsible for the invention, not Monsieur Lepaute.

How did this happen?

First, young Beaumarchais sent five little boxes to the committee. The members of the highly regarded committee peeked into them. After that, they granted Beaumarchais the victory.

What was contained in those five boxes?

Evidence of all the necessary mistakes he had to make in order to arrive at the correct solution.

I MIGHT HAVE THOUGHT that the Cube's structural function consisted of nothing more than holding together the middles; but, in fact, this was the key to keeping *every* piece in its proper place. I needed to make sure that the pairs of middle ones were connected in a way that not only pulled them toward each other but at the same time pulled them toward the middle point. It was like a center of gravity that organized the combined components of the entire structure so that every piece's spatial position was defined and fixed.

Handling the object was really an extraordinary sensation. At last, I had arrived at the moment in which each component was interrelated, interacting with each other: the middle cubes held the edges, the edges held the corners, the corners helped to fix the edges, the edges fixed the middle ones. Its pieces were made of rigid material, but in this combination, they behaved like an elastic ball. I had never before experienced such a combination of rigidity and flexibility, such hard-edged smoothness. But there was more: It invited you to sit and explore it, to start a dialogue with it, to move the elements around from all directions, simply in order to enjoy the tactile experience of holding movement in your hands. Even in its immobility, it tempted you to do something with it, to feel its self-contained suppleness and discipline.

Sure, I may have found the ideal geometrical form, but I certainly did not arrive at the final construction. Now that it was within reach, it still seemed to be far away, in an endless distance, like a mirage. Geometry is based on absolute precision and exactitude: edges should be one-dimensional, equal sides should be perfectly equal, a right angle should be exactly ninety degrees, etc. A physical object can only *approach* perfection of measurement. If the object were intended for mass production, it must be designed to work in spite of slight variations. I don't mean that once I was handling it, I immediately had visions of factories churning out what I had made. But part of the discipline of my profession was to consider the transition from designer's studio to the market. To solve that challenge, I rounded all of the edges inside. There was a mechanical reason for rounding the outside edges: I wanted to make it more comfortable to hold and manipulate without sharp edges that can nick the skin and hurt someone's hands. I rounded the outside edges simply because it looked better.

So what was I doing? Step by step, turning a hazy concept into a very real object, which has then become the very concept of the object.

Making it smooth, making it round, making it comfortable all made it possible for anyone who encountered my creation to experience a light and even abstract concept in its hard, objectified material state. What once was an idealized version of a concept had now found its expression in something that was very real. We were in a kind of hall of mirrors.

If the Cube's joints were fit too closely, it would not have worked well, because friction would have impeded its movement.

In a way, I needed to ruin perfection in order for it to be really perfect.

The image of a spatial perfection had to be imperfect to be perfect.

When you look at it, it is misleadingly simple. All its inherent qualities are hidden; it is a riddle that answers your question only if you make the ride.

On the other hand, if I built it with loosely fitting components, it would rattle unpleasantly. It would not fulfill the goal of constructing an object that, although it consisted of many pieces, formed a homogeneous and closed unit. I spent hours rounding off the edges of the small cubes: since each one had 12 edges, I had to deal with 312 edges. In fact, that meant I had to deal with even more, since after I had done the first modifications, I discovered smaller more complicated details that had to be addressed. It was a boring, monotonous, and exhausting job. In a way it was my version of my father changing the shoreline. But it had to be done.

In the final version, I used spring-loaded screws that pulled the face-center pieces, under constant tension, toward the center. These pieces gripped the edge pieces, and these gripped the corner pieces. A sort of surface tension arose, a round capillary formation similar to that of the water drop floating in a state of weightlessness. I had great hopes for this simple mechanism and was sure that it would work.

At first glance I was not doing anything theoretical. I was just solving some intriguing technical problem, where my hands and eyes served as helpers. I went step by step; when I took care of one technical problem, another one arose. It was too heavy. It was too light. The pieces got stuck. It was something different each time.

But the end result in itself was not a technical achievement, but something else.

You don't think about technical aspects when you twist a Cube. You just want to play. You just want to master it. And there it is, an object that has forgotten its past, like someone who awakens and cannot recall his dream.

I REMEMBER THE MOMENT WHEN I LIFTED the final object off the table and very cautiously began to make a turn. It worked almost by itself. This was the moment I had been waiting for and I did manage to relish it—albeit briefly. Because then I realized, like all newborns, the Cube was naked. Without adornment, all his important information remained inaccessible. The visible surfaces of all the elements appeared to be identical. If the individual elements were not recognizable, it would have been impossible to follow all the amazing moves and see the Cube's vast potential. How could anyone see a change of order if all of the parts looked the same?

I HAD FIGURED OUT how to break the form by turning and bringing it back to the original after ninety degrees. But I didn't see what had changed. In order to make that visible, I had to mark all the elements and give them each an identity.

How about painting each face a different color? Six strong colors made each of the faces distinct, and even the individual pieces became unique. The centerpiece of each face had one of the six colors, each edge piece had two colors, and each corner piece three colors. The combination of colors on any piece was different from that of any other piece, and the colors of a piece told us its correct position in the ordered arrangement. This may appear at first glance to provide some guidance for the next steps, but all this information was not evident for those who were just starting their relationship with the Cube.

My final choices were based on the knowledge and experience with colors I'd gained in art school. I began with the primary colors—that is yellow, blue, and red—and placed stickers of those colors on three neighboring sides accordingly. I then added their complementary colors of green and orange. I settled on painting the sixth face white. I didn't use purple, because, for me, it did not fit with the Cube's masculine character.

I also wanted to make the whole image lighter, so I was looking for strong contrasts. The aesthetic homogeneity of the object—the simplicity of the coloring—seemed important.

Since that time, many different suggestions for colors have been made by all kinds of helping spirits according to some other principles. I am convinced that none of these suggestions, although some did sound interesting, improved upon my original version. The colors had two functions: to identify each of the sides and to give information about the place of the individual parts in the solved state. That is, their noninterchangeable positions. The fixed position in space and time. The fixed point in the Cube universe. At this point, the background color for the Cube's entire body, the color on which all the other colors would be placed, was still an open question.

After painting the Cube, I wanted to be able to follow how each piece moved and track its relationship to one another. I thought it would be easy. I would just have to memorize what I saw. First, I made two turns, and was intrigued to see how a twist changed everything. After two more turns around different axes, I still found it easy to return to its original state. But then, it was like the sometimes exhilarating, sometimes exasperating experience of getting lost in a foreign city. We might go a few blocks and be able to retrace our steps easily. But then we go one or two blocks more, turn to the left or to the right, and soon the starting point becomes more out of reach, until after we walk still farther, and returning to our familiar hotel appears to be almost hopeless. There is no doubt about it: We are lost. There is an episode in Marcel Proust's *In Search of Lost Time*, in which the father impresses his son with the importance of arriving back to exactly the same spot as where they started (which is their house) after a long, confusing walk through the fields, parks, and forests. A journey during which the narrator feels he is far, far away from his home.

If we refuse to accept the fact that we are lost, we become even more confused as we venture further and further from the point where we began.

WHEN YOU TAKE THE CUBE IN YOUR HANDS, there are eighteen different moves you can choose in order to create a new arrangement, by making one or two or three quarter turns (90 degrees or 180 degrees or 270 degrees) with all the six sides (6 x 3 = 18). The fourth quarter turn does not create any change and brings you back to where you started.

Imagine a room with eighteen doors. Which door would you choose? Whichever, it doesn't really matter. After you have chosen one, you reach a similar room with another eighteen doors. There is not only one way from room to room, but when we arrive in the new room, we are again faced with seventeen more doors, an experience that can be repeated infinitely. But after passing through only a few of those doors, you get totally lost.

Back to the Cube: After three turns, it was already a little bit harder but still possible to find my way back on the same route. But after four or five turns, things started to get interesting. Even confusing. I found that it was incredibly difficult to return back to the base. Because each time I thought I was getting closer, I was only getting further away from my destination. To put one face in order was not that difficult: It involved being able to think a few moves ahead, as in chess. The real difficulties began when I tried to continue with the other faces. I wanted to go step by step, in the traditional problem-solving tactic, and quickly realized that this approach simply wouldn't work.

The most interesting point: I couldn't make progress in one area without ruining my progress in another. Which is also a fundamental experience: *Constructing something often starts with destroying something else.*

Not only did I have to deal with the three-dimensional character of all the movements, there were so many different events happening at the same time. It was impossible to follow. I couldn't keep my eye on all the components simultaneously. This was the unprecedented challenge of a three-dimensional object with three-dimensional movement. No matter what one does, one can only see three of the Cube's faces at a time, so in order to map its color territories, I had to rely on my memory.

There was no way back.

I had defined my original problem very clearly. The twenty-six external small cubes were held together by the center cross and they could move freely.

At first, I enjoyed watching what was happening. It was wonderful to see how, after only a few turns, the colors became mixed, apparently at random. But as any tourist on a pleasant or even exotic excursion discovers, after spending some time looking at the scenery, one eventually just wants to go home. After having my fill of the bright-hued confusions, I had enough and just wanted to put the cubes back in order. And it was at that moment that I realized I had no idea what was going on. None whatsoever.

What did I need to do to return to its original position? The door that seemed open was locked with a rusty key that someone had thrown away a long time ago. Or rather, the key was thrown away into the future.

Naturally, it had to be possible to get back to the starting point. At least in theory, just reversing the moves would unscramble it—right? But I soon discovered my limitations, because after moving it randomly five times, the reversal became almost impossible.

At this point, I should note that it had never once occurred to me that I was creating a puzzle. My search for a solution was an end in itself. I was working on an idea to solve a problem that I had formulated for myself—a rather innocent activity without terms and categories. If one is deeply immersed in an activity, really focused on solving a problem or building a project, it doesn't need a descriptive name. After you have a result, you may realize something. When you are in the middle of it, there is no such point. During the slow, sometimes painful, riddle-ridden, or difficult process, there is no definite answer. In the end, when you have accomplished something, there is the experience that it is complete. I solved it. I can sleep. After that, what do you do? Usually, you put it on the shelf. Sometimes you feel you can go further.

The moment of helplessness is the first moment of creation.

I had created chaos and was helpless in trying to figure out how to find my way back. There was no background

information. Nil. There was nowhere I could look to find a solution. I was closed in an escape room I myself had created, and the rules were not written on the wall. How silly of me! How could I have imagined that it wouldn't be a problem to put it back in order again after I had scrambled the colors? I was the first person in history to face a scrambled cube, and as anyone else who has ever faced one will attest, unscrambling it was not going to be simple or quick.

I had a fuzzy, indefinite feeling that if I only did this or that, tried one path or another one, I'd be all right. But the more I twisted and turned, the more uncertain I became. It gradually dawned on me that I had been spending much more time trying to get back to where I had started than I'd taken to get lost in the first place.

I SELDOM DREAM and rarely have bad dreams. But one that recurs is that I am in a strange city and cannot find my hotel. I walk through unfamiliar streets for what seems like hours, getting increasingly agitated and worried that I will never figure out my way back. Getting lost can happen to anybody. We wander and wander, still sustained by the hope that the situation will be only temporary, but becoming gradually more and more uncertain of our ability to emerge whole and unscathed. And then it dawns on us: We've spent more time trying to extricate ourselves than we did getting lost in the first place—even though the distances, at least in theory, should be the same.

And so it was with that first scrambled Cube: I found myself in a totally unfamiliar landscape. I had to solve all the problems that would never have existed if I hadn't created the Cube. Or created the possibility for them to demonstrate themselves. The colors were now so confused that whatever thrill of accomplishment I may have had in assembling it in the first place became utter discouragement. It was as if I were staring blankly at a secret code, which I myself had created but could not penetrate.

This state of being lost usually has one cause: We don't have a clear view of the whole terrain. In the forest, the trees obstruct our perspective. You don't see the forest for the trees, as the old saying goes. In a city, some buildings often prevent us from seeing where we need to go. In our personal lives, a painful problem may fill up our entire field of vision, blocking out perspective and broader context. In the case of the Cube, I held it in my hand. But this was hardly an unobstructed view, because I could not possibly see all of the sides at once. Like an architect walking around a building, the complete perspective is forever out of reach.

In the more than forty years that have elapsed since that very moment, this same experience has probably been shared by many hundreds of millions of people. And like them, I had absolutely no idea how close I was to my target. But there was some hope. Logic dictated that there must be some way to arrive at the solution.

MOST PEOPLE SHARE my experience of being lost during their first encounter with the Cube, but there it is, lurking in the distance, like an unseen animal behind the rustling leaves, or is it the wind? Maybe there it is, in the twilight. We are heartened by the certainty that there is a way to crack this riddle. For me, at that point, that goal was by no means obvious.

I could not figure out if I found it so difficult to impose order on disorder because of my own personal limitations, the lack of perseverance or intelligence, or because there was something here that was so disordered, it might not be solvable by human

intelligence alone. Of course, at another level, I instantly knew this was preposterous.

We all know how frustrating it is not to be able to see something even when all the clues are right there, in full view before us. And when at last we find a familiar neighborhood, we are reluctant to leave it—perhaps it, too, will become inaccessible. Still, the antidote for the fear is surrendering to the experience of being lost. This persistent and recurring state of anxiety is a fresh opportunity to ask totally new and different questions, broader and more challenging ones. Maybe our ancient hunter instincts are revived when all our senses sharpen, as they do when we are in a state of being physically lost. We search more attentively, we listen with greater awareness, perhaps we even detect a familiar scent that we attempt to follow. The first and foremost attitude is to put a stop to any anxiety, because barring the most extraordinary circumstance—a blizzard, a storm at sea—most people are not lost forever.

I couldn't find any true direction with the Cube, no compass to offer some guidance. Even so, I knew the important question was not whether it was possible to solve it. The question was, is it *possible* to find a *method* to do it?

HOW MYSTERIOUS IT IS FOR AN ADULT to observe children. Why is it easier for children to accomplish what we adults find difficult, even hopeless? Finding the Cube's solution is technically easier if someone already has a certain understanding of some mathematical and geometrical principles. The beautiful thing, though, is that knowing these principles is not at all necessary. The Cube can provide an introduction to that way of thinking if the person who picks it up is open to observing, discovering, and exploring these principles. This is why children are often better at solving it than their parents are.

Learning to solve the Cube is not a question of one or two bite-size revelations—whole tomes, largely unsuccessful, have been written on the subject. There are countless YouTube videos walking the viewer through easy solutions. Online, it is possible to find the step-by-step approach. But these all miss the point: To solve it, you must grapple with the very principles that underlie it. And you must also determine your own relationship to it. It is a form of introspection, even if not consciously so. To contrive your personal approach to things, a measure of patience, perseverance, and curiosity is essential.

WHEN I WAS FACED THAT FIRST TIME with the challenge of making some sense out of the confusion of colors, I realized I needed to be absolutely aware of what seemed to be small insights. The nuances. The seemingly insignificant details often have huge significance.

That the middle cubes do not move, for instance, may seem like a trivial discovery, but when everything is turning around an axis, it was useful to see a point of stability. Then I wondered if it would be possible to find the true north by choosing one of the colors as a compass. Over time, a lot of people have found that the best strategy is to use white for help.

One reason, I think, is because white is the only non-color in a strict sense. Often, people have difficulty in choosing between yellow and orange, or orange and red. White is a color that draws us (it is not by chance that the logo is on the white side). I love white. It is the color associated with purity, white is the dress of the bride, and, in some cultures, it is the symbol of death. For me it is the symbol of light and the origin of life.

The point is that it is all about human orientation, our need for some kind of fixed point from which to begin.

In a theoretical sense, there is no hierarchy among the colors. In human perception, one is necessary in order to have a clear starting point. When you hold the Cube, it is easy to twist and turn it and be lost. This is why it is so valuable to have one stable point. There is a saying from Archimedes that with a single stable point, you could turn the whole world: "Give me a place to stand and I will move the earth." So too, with the Cube, it is possible to have a fixed point as orientation.

SOME PEOPLE THINK things through by writing. Through writing, they get a clearer image of what might be going on, and even an ability to predict how things may unfold. I don't. I recognize that this is a valuable technique for making sense of the world; human history attests to it. But here's the rub: The minds of writers work in abstractions; mine doesn't. I am a concrete thinker and an intuitive one. When it comes to the Cube, some people can work out solutions through algorithms without even touching the puzzle. In contrast, I followed my *intuition*.

Intuition is a force that doesn't push you but instead pulls you, draws you toward something singularly important. You can call it inspiration, you can call it anticipation, you can call it a kind of epiphany. An almost mystical state of mind. As lost as I was in figuring out how to bring order into the chaos I had created, I also experience great joy, almost a kind of trance while working with my hands, actually shaping things, handling materials, creating tactile forms, experiencing the process of discerning beauty that is locked in difficulties.

In 2010, long after the Cube had become famous and millions of people had solved a thing once so perplexing for me, I read a story about a thirteen-year-old speedcuber from upstate New York. He had just won a big cubing competition at MIT and insisted that he could teach anyone the solution in ninety minutes. "People think it's spatial recognition, math and intuition, but there's really no math involved," he said. "I just think about the cube in layers or in chunks of bigger pieces and then I can do it. The first couple of layers are intuitive, and then I just use a sequence of moves I've used before." I was impressed by his candor when he talked about the way he operated intuitively to solve the early moves.

For me, intuition is the process of noticing something and seeing its significance beyond the phenomenon. The experience is one of not knowing but feeling some power that pulls you.

"CUBE SOLVERS" can be divided into two camps: those who rely predominantly on intuition and those who mostly use algorithms. The intuitive folks tend to be neophytes, and they are often quite young children or older adults. They struggle, they look, they often have a keen sense of design, and they, too, "think with their hands."

One cannot run entirely on intuition, of course, and some experts consider those who do to be ultimately doomed.

Then there is another group—especially teenagers and young adults—who have mastered a variety of well-founded and reliable algorithms that provide paths to solutions. I have discovered a vast parallel universe of YouTube videos, Wiki groups, and online cubing communities in many languages that offer algorithmic instructions for solving my often rebellious and wayward offspring. What is so beautiful is that there isn't a single answer—which is true about so many or all things in life—but an inherently rich, cascading series of moves that are interdependent with other moves.

When attempting to solve the Cube, I never once thought about it in terms of algorithms. I simply paid attention to only a few of the pieces and ignored what was going on with all the others. I just wanted to grasp the result of the series of three-dimensional movements using different axes. That was already difficult enough. I followed my instincts, I thought with my hands, I used the full force of my intuition as I worked the corners. I figured out how to liberate pieces that I had blocked and to manage the frustration of having pieces that I thought were in the correct position turn out not to be oriented properly.

AT FIRST, I PUT FOUR CORNERS in their correct place. This can be done intuitively; you just need to understand what "correct place" means. If you start with the white side, as most people tend to do, the corner pieces are white on their top and their sides match the colors of the neighboring middle pieces in the middle layer. Once this is accomplished, holding the Cube with the white side up I established the "top layer," which meant I had some orientation.

But let me clarify what I mean by "orientation" here.

At its most basic, orientation in the Cube world means the relationships of the small cubes, the cubies to the middles, which is known in the Cube world as "the 3D cross."

If there is no orientation, it is like something that is weightlessly levitating in outer space. You don't know if it is upside down, or lying down, or standing up, because there is nothing with which to get it oriented. But each of us are in the center of our own universe. When we stand, we know the Earth is below us, the Sun is above us and we are surrounded by many things in front of us, behind us, to our left, and to our right. We knew that the cold comes from the North, warmth from South, the sun rises in the East and sets in the West. We take all of this for granted; we are at home.

But not in the Cube world, which is strange, unfamiliar, unusual, governed by its own rules.

I came to understand that one of the essential challenges in solving the Cube was to discover the nature of the orientation of the cubies. I found that there are two components of it; each one has a place of its own in the 3x3x3 grid, and in this place there is only one correct position. The edges have two possibilities, and the corners have three.

The next phase of solving my very first Cube was to place the remaining four corners correctly on the opposite (i.e., bottom) layer. The corner pieces were already in the lower layer but not in their dedicated place. They did not match the yellow middle piece on the bottom nor the green, red, blue, or orange middle pieces on the sides.

Here things get a bit tricky, as you have to find a series of movements that change only what you actually want to change while keeping the rest in place. Everyone who has experienced the Cube can sympathize; whenever I put one of the pieces in order, I destroyed the order of all the others! My challenge was to find the precise series of movements which would not result in all of the elements changing places or positions. If I could find two or three elements to remain fixed while the others moved, that would constitute great progress.

The whole process reminds me of the observation from the old Greek philosopher Heraclitus: "If you do not expect the unexpected, you will not find it."

Gradually, I found a series of movements where I could ex-change the places of two corners, while the other six remained stable. I was not just seeking to find a series of movements; I was attempting to find ones that were easy to remember.

I know it must seem quite strange, now, to imagine that getting only the corners of a 3x3x3 cube in the right order would take weeks, when today some preschoolers are perfectly able to figure this out in a matter of minutes.

But remember: This had never been done before.

The whole notion of a scrambled Cube, of a puzzle, like this, that had a clear solution, had not yet entered anyone's consciousness. After some persistent work—fitting it in after teaching and socializing and preparing for classes—I located all the corners of the 3x3x3 in the right place and the right position. All that was missing were the edges.

Solving the 3x3x3 turned out to be a very complicated task. However, I realized that I first needed to solve the 2x2x2, which was the foundation for the 3x3x3! In fact, the 2x2x2 cube equals the eight corners of the 3x3x3—it just doesn't have a middle layer. You can choose any of the orientations of the 2x2x2, but everything seems equal, since there are only corners and no middles. You need to choose one of them for orientation. The one with the primary colors of red, blue, and yellow, for example. The world record for solving the 2x2x2 was achieved in 2016 by a then thirteen-year-old Polish boy named Maciej Czapiewski, who solved it in less than a second—0.49 of a second, to be precise.

But, for me, the Cube, being what it is, still insisted that a great deal remained to be resolved. I still had not conquered the edges.

WHAT ARE THE REASONS IT IS often difficult to make a good decision in life? Often, it is the lack of sufficient information, so therefore collecting more information can help lead to a better decision. With the Cube, this is not the case: All the information is readily available, nothing is hidden. The other difficulty in making the right decision, which is probably more common nowadays, is too much information: Sometimes the information just complicates matters; other times it is simply false. But again, this isn't the case here. *The Cube never lies.* And all the required information we need is right there. Nothing more, nothing less.

So, then, what is our problem? It is the difficulty in finding only the information required for the immediate task at hand. You need to select one move among many, and that is far from easy. Especially doing so consciously. In everyday life, we are constantly and unconsciously engaged in 3D movements, with our brain guiding us after having engaged in an endless number of experiments in the past. We don't pay attention to how we walk, only where to go. Speedcubers achieve this level with the Cube by practicing for huge amounts of time.

Based on these early experiences of mine, I would suggest that a beginner should start out by hiding what's not important. You could use black stickers and cover the whole Cube in black, except for the middle pieces. Then take off the stickers from only two of the neighboring corners. Then do a few random turns, and try to move them back next to each other again. After a while, you start understanding the results of the series of turns by using different axes, and your fingers will also remember those moves. Then you can uncover the next few faces: first uncover the three exposed sides of the corners, then the edges. Eventually, you will get back to playing with your fully colored and solved Cube!

After you are able to solve the 2x2x2—which means that all the corners are in place—finding the way to put the edges in their correct places is much easier. It becomes possible to do it mostly by intuition. When I found an algorithm to make the "flip," turning two edges 180 degrees simultaneously, in their place, without changing anything else—I finished my battle for the solution with a victory.

BACK TO THE STARTING POINT. So once more you can get a sense of some of the reasons why I hate to write so much. Because even though you can find long descriptions of the algorithms to solve the Cube, in plain language it is practically impossible to describe the kinds of movements and appearances of the Cube *without handling it and seeing it*. Our mind is not prepared to cope with something it has never before seen or done, especially with the problems that three dimensions give to us. As I have already said, no one can ever see all six sides of the Cube at the same time (without using a mirror behind it). And even when being able to turn it, it is impossible to follow what that means without watching it in action.

When we want to turn a table around in a room, we simply lift it up and turn it around. But in the case of the Cube, *I had to turn the room around instead.*

One has to understand what is constant and what is changing. Generally speaking, we do what is simple to do. If I move a whole stripe in any direction (three adjoining cubies), in fact, it stands still: The relationship among the cubies (or colors) is constant. Small discoveries, the fractions of solutions, took place incrementally. We all know the experience of losing something—a car key or some money, your spectacles or your passport—and then suddenly finding it, simply by retracing your steps, going back in time, reproducing the actions you did without consciously registering them at the time. But it is a totally different emotional experience if you lose a pair of something—gloves, or socks, for example, or earrings. Finding one half of the pair offers some sense of relief and some limited satisfaction, but we are immensely frustrated by the lingering knowledge of its lack of completion, of its being a fragment. Eventually, the discovery of its mate is accompanied by the strong feeling of accomplishment. That was exactly what it was like for me. When I found a single element of the solution, I understood how it worked and what its place or function was in the whole. By repeating elements of movement, I could achieve my objective.

And then, finally, in a wonderful and memorable moment, it all fell into place. In the end, it took me a full month to get back to my starting point.

I looked at it, and all the colors were where they should be. What a fascinating feeling this was! The mixture of a great sense of accomplishment and utter relief. And a real sense of curiosity: What was it going to be like to do it again? What other discoveries would I make in the process? What did I learn about this crazy Cube's essential nature?

I have read some accounts of my discovering the first solution, and most of the time it is depicted as if I were living day and night with the Cube, locked in my room and obsessed with solving this obstinate object. But in reality it was just a pastime. Time flew. I went to work. Spent some time with friends. Lived my life. Kept up my daily routine, and in the meantime, I was perfecting the structure, making new models that worked better. When I had some time, I would play with solving it. Eventually, things got more involved, I got quicker at solving it, and the Cube began to dominate my life.

But that was a bit later.

APPROXIMATELY SIX MONTHS AFTER delivering my brainchild, I realized that it was time for me to move to the next level, to figure out some way to manufacture it, because clearly, I was not the only person who found it interesting. Friends and acquaintances I showed it to were as struck as I was by its charisma, and my fascination with this object made me wish that others, too, could have access to it.

My difficulty, and the difficulties experienced by others to whom I gave it, revealed that the challenges presented by the Cube were not trivial at all. If I can't do it, it's not because I am stupid, but because it's an extremely complex problem. This is an intellectual pleasure and what, in the end, we call a good puzzle. A puzzle that is embodied by an object: Take it apart, or put it together! The puzzle itself is contained within an object (for example, as opposed to being given four matches and told to make as many triangles as possible using all of them). If you solve the Cube, it is done. There is no reason other than just to prove to yourself that you are capable of beating the challenge, of not being daunted by the difficulties of the problem, the strange nature of the problem, the complexity of the problem. When I was working on this and thinking about it, I realized the problems of the Cube that demanded to be solved had many more dimensions than I had realized before.

It was only when my friends looked at it and tried to figure it out that it occurred to me that what I had made was more than a tool for the theoretical purpose of illustrating spatial moves. It might even have some commercial potential. At the time, no one pointed this out to me; there were no entrepreneurial friends noting its potential in the marketplace. My decision was intuitive, instinctive, based on my sense that the Cube seemed new and original. And since I considered myself to be an average person and liked it, then others would too. We average people are in the majority, after all. Obviously, I performed no market studies, but I did have some knowledge about what puzzles were out there, and I saw some potential.

WHAT I DIDN'T HAVE AT THAT POINT was some fantastic dream of giving up everything, stopping my work in the Academy, making a 180-degree turn, and planning to spend the rest of my life creating fantastic, incredible, irresistible puzzles. My vision was relatively small. I had something in my hand that was quite interesting, I thought it would not be complicated to manufacture. So why not? We could have a go at it.

Since the Cube was a new concept, I realized that a basic first step was to protect it and potential future partners by getting a patent. That is how things go. So I went to a patent lawyer. The patent lawyers aren't exactly scientists, but they are just as meticulous.

The patent lawyer had me make drawings and write an accompanying text to describe what I did. You can imagine how eager I was to begin writing about this. I hated the task, and so it took several months. I created the Cube in the spring of 1974, and on January 30, 1975, I submitted an application to the Hungarian Patent Office. In it, I described the Cube as a "three-dimensional logical toy."

At that time Hungary was a socialist state economy, but some limited free enterprise had slowly begun, with some small shops, private restaurants, and family farms. The real entrepreneurs were the proprietors of industrial cooperatives, where workers were all partial owners. Patents were for technocrats or mechanical, chemical, or electrical engineers—not artists. Architecture and design were considered closer to the humanities, and just as one would not patent a poem or a musical composition or a painting, one would not, in most cases, patent a building or a well-functioning object.

But I had some knowledge of the importance of patents because of my father.

Some of my father's designs had been patented, but his work was not rewarded at the level of its importance to the field. When I was young, my father had been involved in a lawsuit concerning a patent fee. He had sued the factory over one detail involved in the construction of one of his gliders. My father was in a difficult position because his designs all had to do with flight, with new designs of gliders and wings. Here comes the tricky part. Hungary was a member of the Warsaw Pact, and in that system, each country was restricted to only manufacturing certain products and was proscribed from manufacturing others. Given that the aircraft industry potentially had military importance, the Soviet Union kept very strict control over that sector.

At the same time, the gliders that my father specialized in, and for whose designs he had several patents, were crucial for the training of pilots. There was one detail that he had patented involving only a very small part of the overall construction that he had created—which of course was much more significant and valuable than the patented detail under dispute—but he felt it was worth fighting for. He wanted to point out to those who thought that there was no problem in separating the idea from its originator that they were wrong. He won in court, but his victory was more symbolic than monetary. Still, symbols do matter.

At last, on October 28, 1976, the Hungarian patent for the Cube was published and registered. During the whole process, the only question that was posed to me was how my toy was distinct from a French toy that was already patented. The answer was simple: The only characteristic they had in common was that they were both three-dimensional. (The French toy was a variation of the Chinese Cross.) The similarities ended there. At long last the announcement appeared in the *Hungarian Patent Gazette*, and later the official document confirming the registration arrived. It was tied with a red-white-and-green ribbon, the colors of the Hungarian flag, and stamped with an impressive wax seal. I felt some satisfaction. *Here was the proof that I had made something nobody had ever made before.*

While I was waiting for the patent, I needed to find a way to manufacture my creation in a small country with no particular affinity for toy production. Industrial cooperatives allocated a negligible fraction of their production capacities to manufacture board games, balls, dolls, and simple toys for the home market and for exports in small quantities.

One of the fundamental lessons one learns as an architect is the importance of materials, how they function, how they respond to stresses and the environment.

In the workshops at the Academy, I worked with a few friends who helped me test different materials, which naturally required different tools. There are peculiar kinds of plastic, and some are so strong, you need professional equipment that is quite similar to what one uses when working with metal. And it gives very precise results. Had I come up with the idea today, with 3D printers, it would have been a very different process—and certainly a faster one. At first I was working with hard rubber, which wasn't elastic at all in the way we are accustomed to imagining a rubber band, for example. This was strong and sturdy and black. I then turned to plastic as the material, and the process needed to make the actual cube was something called *injection molding*—which is exactly as it sounds, the process of injecting a molten material, in this case plastic, into some kind of cavity that is a particular shape. It was a technology capable of high-production yields and pinpoint accuracy. I knew this would be able to produce a high-quality, affordable Cube. But it didn't move the way that plastic did, so finally, with the help of some friends in different workshops, I made my first few prototypes. I finally came up with something that was light, resilient, inexpensive, and easy to shape. I knew that its very simplicity meant that it would be easy to manufacture. When looking for manufacturers, I brought my working models to them.

SIMPLICITY AND THE EASE OF MANUFACTURING are closely related concepts for me: What is simple will be easy to manufacture. I also wanted to improve the construction to the degree that it worked perfectly. The number of potential faults would need to be reduced to a minimum. Making the Cube simple would also ensure it was inexpensive. Its production would be profitable for the manufacturer and the trade, and remain affordable for the consumer.

Basically, my design aesthetic errs on the side of simplicity. Objects should have nothing redundant and not say more about themselves than what they are: The object is truthful. I didn't make the constructional rules of the Cube. They are natural rules; I only recognized them.

Once I had decided on plastic, I made models from white plastic, from black plastic, all before I found a manufacturer. With luck, it took me only a short time to find a cooperative that had the capacity for injection molding on a modest level. They had manufactured chess sets and plastic toys for the Hungarian market.

I didn't just give them the design and hope for the best. I am very exacting—I suppose this is another thing I have in common with my father—and I felt it was imperative that my instructions were clear and unambiguous. There were many decisions that I had to make. What would the optimal thickness of the material be? Which plastic would be best? Manufacturing plastics was a field of its own. I knew that at a certain point I would need to leave the execution to the experts. My greatest concern was how the end result would be affected, especially because geometrical exactitude is a key characteristic of the Cube.

This applies not only to precision of measurements, but to its precisely defined form as well. Its parallel faces are exactly parallel. But the very nature of injection molding prefers tapered forms. I did not attempt to make that technology do something it could not do, but I fought for reducing the irregular aspects to the point when the irregular did not appear to be so. If you examine your Cube thoroughly, and hold it against the light, you will notice that there are tiny gaps in the adjacent cubes. These tiny gaps show the two touching surfaces are not perfect—they can never be—and in that sense the light can partly go through it.

Another basic feature of the Cube is its solidity.

By this I don't mean that it has to be actually solid, but that it should give that impression. Solidity and a closed formation are akin for me. Although we know that both a ball and a jar are hollow inside, balls appear more substantial compared with the emptiness of a jar. To accomplish this impression of solidity, it was important that the colors and the construction of the Cube didn't look flimsy in any way. Complying with my demand required an additional element in the manufacturing process. The little top for closing the cavity had to be manufactured, and a special tool had to be made for producing it.

Weight is another factor that affects the impression of solidity. Things that are as light as breeze do not evoke a feeling of solidity or stability in us. That is how the human psyche and perception work. But clearly it needed to be light enough so that it was easy to play with no matter how young or old the cuber might be.

It was important to avoid the impression of something that was cheap and weak. The material one uses for the product is very important. And the first series of Cubes were surprisingly heavy. The subsequent ones were somewhat lighter, but Cubes have always been heavier than other objects we are used to in toys made of this size and of plastic. We have all sorts of expectations of what something will feel like when we look at it—its appearance and our experience of objects made of similar material will make us believe that we have a good sense of what it will be like when we pick it up. We have all seen toys that are a bit like tricks: the plastic brick that looks as if it should have weight, but when you pick it up it is featherlight. Or a child's black dumbbell that gives the impression of something one would find at the gym weighing fifty pounds but again weighs nothing. The Cube may have appeared to be very light and composed of hollow little elements, but, in fact, it is more substantial than people might have expected.

Another manufacturing detail was connected to the formal perfection of the Cube. One of its important geometrical characteristics is that its faces are planes. Any detail that breaks or disturbs this plane disturbs this image. In the first manufactured series, because of the irregularities in the fitting of the two halves of the mold, the center line of the edge was a bit uneven, as were the contours of the small cavity surrounding the middle cube, which was the last element fitted in the inner construction. These small blemishes were covered by the colored labels, but even so, they affected the appearance—at least for me.

The small flaws may be hidden, but if the object is not perfect, it cries out, and sooner or later external flaws can modify the inner image of the thing. The inside must be handled with the same scrupulous aesthetic care as the outside.

Then there were important issues about the screws and springs that held the Cube together. In mass production, we had to find the appropriate force—the precise springs, screws, and method of assembling—which would ensure that the force holding everything together should be no more and no less than required. The key was to make it so that it would not fall apart and yet be easy to handle. Perhaps these details don't seem to be that important, but they were and still are for me. I wanted this object, as a product, to be as precise as a scientific instrument or as carefully, perfectly executed as an object made by a nonfigurative sculptor. My expectations came partly from my belief that just because the Cube was a toy, that did not mean it was inferior or that this could be used to justify shoddy construction.

When manufacturing a product, one needs to know what kind of demand to expect for it and the magnitude and pace of that demand. For new products, it is of course done by quite rough estimates, and the newer and the more original the product, the more unreliable the estimate. It was in order to get an idea about this that I went to the largest Hungarian wholesale trading company for their help in arriving at the best potential market estimate. In the Hungarian toy market, they said, ten to fifteen thousand units sold would be considered a great success. They estimated that the Cube would sell about ten thousand pieces in the first year, but gave the co-op an order for just five thousand Cubes to start.

4

It's still magic even if you know how it's done.

—TERRY PRATCHETT

NEARLY THREE YEARS after I submitted my patent application, at the end of the year 1977, a product in a simple blue box, named *Bűvös kocka*, or "Magic Cube," appeared in Hungarian toy shops. Naming my creation was not difficult for me. In the beginning it never occurred to me to attach my name to it—that only came later—so I tried to be both literal and suggestive; yes, it was obviously a geometrical shape, but it was more than that. In a natural way, it came from the "magic square," the ancient puzzle, in which one had to fill a 3x3 grid with the numbers one to nine in a way that all the sums equaled 15. The magic came from how it drew you in, put you under a spell while you played and struggled with it. I was also thinking of a magician's black box, in which objects disappeared inexplicably only to reappear transformed into pigeons or rabbits.

On the box was a note from me that said:

A toy for children and adults to advance their logical thinking and sense of space. The 26 small colored cubes that are visible can be arranged not by taking them apart, but by turning them into practically infinite positions. All six colors can be mixed in many ways on each face. The objective of the game is to make one face monochrome, which means making a single color on each of the six faces. It is a very difficult problem to arrange several faces simultaneously, and it can be solved only by recognizing the laws governing the turns. To put one face in order within 15-20 minutes is a pretty good result and shows good logical ability. Watch in what fashion the elements change their place after different turns in different directions. The laws you will have discovered this way may be your guides to the solution.

You will note that I said there were twenty-six cubes visible, but in fact, there were twenty-seven pieces. From the beginning, the hidden core in the center remained a mystery.

There was no advertising budget, no publicity campaign. Then, quietly and persistently, the Cube began to find his customers.

Someone bought one, then instantly bought another one to give as a gift. Parents bought one for a birthday, and also puzzle collectors stumbled upon it and found it compelling. Children who received it as a Christmas gift took it to their fathers for solutions (since fathers know *everything*). Then the father got absorbed in it, and the children begged their fathers to let them play with it at all.

FROM 1977 TO 1980, the Cube began an independent life. It attracted friends in Hungary, traveled in parcels and in briefcases. Relatives visiting members of their family abroad or tourists visiting Hungary carried it in their luggage next to other Hungarian delicacies like sausages and Tokaji wine. It traveled in students' backpacks as they hitchhiked throughout Europe. It appeared next to the lecture notes of scientists going abroad to conferences.

In 1978, when my puzzle won a prize at the Budapest International Trade Fair, it was the first "official" recognition that the Cube was something special. An interesting acknowledgment also occurred that year when the Ministry of Culture awarded me its annual prize. It was the first indication of something that had never occurred to me but would turn out to be a powerful aspect of the Cube's identity: the Cube as a cultural signifier.

But these prizes, although gratifying, could not compare with the wonder I felt when seeing how many people were captured by the Cube. By the end of 1979, 300,000 cubes had been sold in Hungary and about 50,000 abroad. With a Hungarian population of only 10 million, sales on this scale were extraordinary. One of a kind. One could easily say that by the end of 1979, the Hungarian population was addicted.

But.

Here comes another but.

Introducing this product to a broader market was neither quick nor easy. The major international toy companies all rejected it. They felt it was too difficult and didn't conform to the commonly held notions of what a puzzle was or what made a toy sell—the solution of it seemed to be too easy, but in fact it was intimidatingly complex. To learn the objective requires less than a minute, but it takes a lifetime to master.

Maybe some of the companies' buyers tried to play with it, got frustrated, and then assumed no one else would like it! Who knows?

It turns out they underestimated the desire for both children and their parents to be challenged. And they didn't realize how addictive it was.

Puzzles are in general a small fraction of the toy market. Prior to the Cube, usually you weren't able to buy them in a toy store; you would find them in places that sold souvenirs instead. A toy company that took itself seriously, the general impression was, did not sell puzzles.

And here's the twist.

Small nations are sometimes great.

When its members are spread out all over the world, a kind of an informal network is created. Hungarians who do not know each other are still connected by our impossible language (seemingly unrelated to any other language) and when we meet abroad, often connections are established that would never have existed back home. Somehow those connections seem deeper than they really are.

I was all alone with the Cube from inception to delivery and from challenge to solution. I was still very much on my own searching for manufacturers in Hungary and persuading state-owned, socialist companies to distribute the final product.

But to make the Cube the global success it eventually became, I fortuitously happened upon a true partner.

Born in Transylvania (then of Hungary, now part of Romania), Tom Kremer survived the Holocaust as a young boy. He escaped to Switzerland and then started a new life in Israel. He later moved to the UK and found a lovely wife from the famous Balfour family.

Tom was a headstrong man with a curious intellect, interested in many things from literature and education to politics and philosophy. Fortunately, his many interests included games and puzzles, so when he decided on making a living in London, he opened a small firm investing in new toys. He named the company Seven Towns, in memory of his birthplace in Transylvania. Seven Towns became a reasonably successful venture, working with inventors in the toys and games area and selling ideas to large toy companies around the world.

Every year since he had set up shop, Tom attended the Nuremberg Toy Fair. He was also there in 1979.

There he overheard a wandering entrepreneur speaking Hungarian—or German with a heavy Hungarian accent. And so, Tom listened and understood, while no one else did.

The entrepreneur was trying unsuccessfully to tickle some interest in an impossible puzzle toy named the Magic Cube, which was apparently wildly popular in far-away, behind the Iron Curtain, Hungary. Serious professionals had no time for him. But the amateur toy inventor from the city of Kolozsvár in Transylvania absolutely loved it at first sight. It was Tom's agility and unwavering belief in the Cube's potential that made it a commercial success internationally—several times over. Finally he persuaded Ideal Toy, a large American company that had sometimes struggled in the past, to put a large last bet on the Cube. This company, which was created by the couple who invented the teddy bear in 1903, appreciated the potential, and signed a contract for 1 million Cubes to be sold overseas. And years later, when sales eventually collapsed following the craze, Tom diligently started to buy the rights to the Cube for Seven Towns. And he waited and waited patiently until we could finally relaunch the brand. Seven Towns was the home of the Rubik's business for over three decades. When Tom retired and his son took over the company, after a while we decided to establish a new firm with its sole focus on Rubik's Brand.

Tom Kremer passed away in 2017, but his phenomenal commercial accomplishment of bringing the Cube to the people lives on.

Back in 1980, when Ideal Toy had just started the Cube's distribution, they needed a marketing plan. And they were not keen on "Magic Cube" as the product's name. When the Cube gained some attention internationally, the name "Magic Cube" would be impossible to trademark, and that was key for its protection. As Shakespeare famously asked, "What's in a name?" But in this case, the answer was quite a bit. So they tried to come up with some other ideas.

These ideas were duly rejected. The best, the most exact word is one that is completely self-referential, but the words "magic" and "cube" appeared on too many toys. At some point, the idea of using *my name* for the product was put forward by Ideal Toy. At the time, I was told that for a name to be accepted as a trademark in the United States, it had to be rare enough that it didn't occur more than ten times in the New York City telephone directory. This turned out to be apocryphal, but I also heard that "Rubik" passed that test. My name just worked on the Cube: It was short and crisp, unusual but not exotic, and easy to pronounce in many different languages. It remains recognizable in any accent, has no associations with notorious personalities, and is not common. There is also a nice one-two rhythm to it, and there is something almost onomatopoeic about the "b"—that suggesting a beat and movement—and the sharp-edged tone of the "k."

Ideal Toy sent me a letter of consent, which I still have somewhere.

For one dollar ($1.00) and other good and valuable consideration, receipt of which is hereby acknowledged, I, Ernő Rubik, hereby give my consent to the use and registration of my name as a trademark by Ideal Toy Corporation, 184-10 Jamaica Avenue, Hollis, New York, 11423, for games, namely puzzles.

AND THINGS TOOK THEIR COURSE. On January 10, 1980, I went to the notary public and signed this document. Even as I understood the importance of naming as an act—after all, at this point I already had one child—I didn't appreciate the full significance at the time. My ignorance about this subject is strange in retrospect, but we human beings are imperfect, which is what makes us perfect. The truth is, without a name, something doesn't really exist for us. Naming an object distinguishes it and also helps us to understand it in relation to something else. I must say, it never occurred to me to attach my name to my Cube. For me, it was just a formality. And I obliged because I did not want to obstruct what appeared to be quite a normal procedure.

As I look back, I might have hesitated if I had thought more about my name appearing on tens of thousands of boxes. But I didn't. Kierkegaard said, "Life can only be understood backwards; but it must be lived forwards."

As with many things that are meant to be, the name of the Cube seems logical, yet again, only in retrospect.

BETWEEN 1980 AND 1983, so many things happened so quickly. During these years, the chain linking me and the Cube drew tighter and tighter. At the same time, launched into the world, he led a much more independent life. He went his own ways, while I chased after him. I was so busy keeping my head above water with all the demands of travel and appearances, that much as I would have loved to describe the details of this period, they blur together in my mind.

I do remember that in January 1980, I received my first blue passport. Because we were behind the Iron Curtain, Hungarians were not permitted to travel freely. Many of us had red passports, which meant we could travel in the fraternal socialist countries. I had already visited a number of eastern European countries by then. In my student days I visited Poland regularly to ski in a resort near Bielsko-Biala. Some summer vacations I went to Bulgarian and Yugoslavian beaches and swam in the Adriatic and in the Black Sea. I knew the Baltic regions from a study trip to East Germany, and a student exchange program once took me to Moscow and Saint Petersburg (then Leningrad). The West, however, was not an option. It required a blue passport, which only a few Hungarians, mostly diplomats, possessed. During my childhood, the opportunity to travel in the West simply didn't exist. When I was a student, things had loosened up a bit, but I could never afford it.

During that time, foreign trade was monopolized by the state. The foreign trade company managed the connections with the Western companies, buying and selling—mostly buying—because it was part of their portfolio. After the contract with Ideal Toy was signed, I needed to travel to the United States and other Western countries to demonstrate the Cube and explain it. So I gave the authorities a photograph and an identity card, and my passport was expedited. As the Cube took off, so did my opportunities for travel. A trip to the Big Apple with my new blue passport was my first journey to the West, my first flight overseas, my first visit to the United States, and my first business trip ever.

It felt like a miracle. Or maybe a fairy tale is more precise.

The purpose of my trip was to attend the American International Toy Fair in New York and introduce what was now being called the Rubik's Cube to the American market. Ideal Toy wanted to launch the product in a significant way, and having the eponymous inventor there—who had such a fascinating, that is, very limited and strongly accented, way of speaking English—seemed a good chance to attract some attention. There are toy fairs all over the world—London, Paris, Valencia, Tokyo, New York, and Nuremberg—and they are all different. Usually, these fairs are for the professionals in the toy business and not for children, who are deprived of the opportunity of seeing thousands of toys in one place.

My job in New York was to reassure people who were interested in the Cube that it was possible to solve it. The toy fair felt like a circus, and I was there as part of the show to demonstrate the magic of solving the Cube. I had been doing it now for close to six years, so I was proficient.

After my first experiments, I had come up with a system: first I solved the corners and paid no attention to anything else. I then solve one layer, any one of them, an ad hoc decision depending on which one I was working on. Next I solve the layer on the opposite side. The last stages for me are the middle area.

This may not be the fastest approach, and even though it sounds very simple, it still involved quite a few steps. To this day I am not keen on doing the research to find some of the shortcuts that are now commonplace. One big reason is because I am lazy. Why should I work hard on something when I have a perfectly adequate way of doing it? Another reason is, I don't have a very good memory. I remember something only if I consider it to be important. I remember faces, but not names, places but not particular cities.

If we sat down now, I could probably solve the Cube in a minute, which is a good time for a competitive beginner, but now kids are cutting it down to seconds. Less than half a minute, a quarter of a minute, now a speedcuber can solve the Cube in less than a tenth of a minute. No one can go to a competition nowadays unless they already can solve the Cube in no more than fifteen seconds.

Back then, during my first trip to New York, which was a time long before children became professionals and semiprofessional Cube solvers themselves, it was not that easy to find anyone to demonstrate that it was possible to solve. (The toy company also probably realized that it would not be the worst kind of advertisement to have the real Rubik accompany his object on his journeys.)

My arrival in New York was a thrill and a bit of a shock. My English was not very good at the time—I still wish it was better—so being surrounded by that language and doing interviews both with the media and with the businesspeople attending the fair was not at all simple. I solved the problem by answering questions I thought they *should* have asked, instead of the ones that they actually did, or I spoke freely about what I thought was important no matter the subject they might have raised. Even after my English improved a little, this remained a winning strategy.

AT THE HUGE TOY FAIR, I positioned myself in the area devoted to the Cube. Sample Cubes were available if anyone wanted one. I never could have imagined the toy business was on such a large scale and was taken so seriously.

Not being able to solve a problem is the best advertisement of the problem. As we know from the famous mathematical problems that needed, or still need, hundreds of years to be adequately solved.

The trip to New York involved a lot of publicity, so that after a few months, the Cube reached the stores and soared into the stratosphere. How did this happen? What forces converged to create such a massive cultural phenomenon? To this day, I cannot say. What I do know is that much of what happened had little to do with me personally—particularly on the business side. The singularity, the ability to appeal to people across generations and cultures, was certainly important. But the Cube's rapid ascent was also powered by the efforts of a handful of people who perceived its potential immediately and believed that the world would too.

Somehow, because there were so many stories about it, and it was so easy to transport, and it appealed to every age with really no distinctions on who would find it compelling (and, to be honest, the fact that there were so many counterfeits resulted in it seeming to be everywhere), the Cube became a craze.

In the first three years after it was licensed to Ideal Toy, 100 million Cubes had sold worldwide. On June 12, 1981, a how-to guide topped the *New York Times* bestseller list, with the title *Mastering Rubik's Cube: The Solution to the 20th Century's Most Amazing Puzzle* by Don Taylor. After three weeks another one joined, *The Simple Solution to Rubik's Cube* by James G. Nourse, which became the bestselling book of 1981, selling more than six million copies that year, and was the fastest-selling title in the thirty-six-year history of Bantam Books. A thirty-six-week streak continued, in which several Cube books were consistently on the list. On January 24, 1982, there were six.

Demand was never fully satisfied and the toy factories were never able to catch up. It appeared on the cover of Scientific American in March and Time in December, along with a torrent of other appearances in articles and on TV.

On the cover of an American book around this time was a hand with a Cube handcuffed to it, the joke being the Cube was addictive enough to make you his prisoner. For me, this joke was a bit more personal. Before then, I had lived my life like a leaf in the breeze, but now I was in a hurricane. In the storm that this phenomenon had whipped up, it was not easy to maintain one's balance. At that time I had no money to speak of; whatever financial benefits I enjoyed came later. I didn't receive a salary. There was no contract that defined how much money I would eventually receive. I was supposed to be paid from the royalties, which were just expectations; it was only a projection. I was still very young at the time, and completely inexperienced in dealing with finances. My basic approach, then and now, was to make sure that I never spent more than I earned. I was happy that my travel was paid and I got some pocket money as a per diem. But even that I didn't need. I had no time to spend money. I had no need to spend money. And I had no money to spend. So life was perfect.

But during the craze, the Cube phenomenon was not connected with somebody but with everybody. This was much bigger than just me as an individual. When I was in the middle of it, in the eye of the hurricane, I didn't realize that I was in the middle of what we would look back on and describe as a "craze." In war, when there is a battle, the soldiers fighting it have no real perspective. It is too close. Only with the distance of time, or perhaps an airplane flying over it, is it possible to get some sense of the event as a whole. I may have been in the middle of the craze, but what exactly was a "craze"?

I had no idea what it all meant, nor did anyone else until it was over. All I knew was that there were demands to go here, there, and everywhere. In July 1982, Douglas Hofstadter wrote a second article about the Cube in *Scientific American*. One passage answered many of the questions that I was often asked. "I would like to close by discussing the astonishing popularity of the Cube," he wrote. "People often ask, 'Why is the Cube so popular? Will it last? Or is it just some sort of fad?' My personal opinion is that it will last. I think that the Cube has some sort of basic, instinctive, 'primordial' appeal. . . . I am confident that the Cube, as well as 'cubes' in general, will flourish. I expect new varieties to appear for a long time to come, and to enrich our lives in many ways. It is gratifying that a toy that so challenges the mind has found such worldwide success."

It seems as if he was an excellent prophet, as he envisioned the future. And I followed the blazing tail of this comet.

Except . . . later that year, it seemed as if Hofstadter's prophecy might have not been entirely correct.

AS SUDDENLY AS IT HAD BEGUN, it was over. One moment it seemed everyone had a Cube or wanted a Cube, wrote about the Cube or filmed the Cube, talked about the Cube or came up with new strategies to solve the Cube. And then, by the end of 1982, it seemed the world had lost interest entirely.

The *New York Times* wrote the official obituary when it announced that "the craze has died" in October 1982. The irony was that, the same year, the very first—and for a long time the very last—official Rubik's Cube World Championship was held in Budapest. It took place on June 5, 1982, and representatives from nineteen countries appeared to compete. Mostly they were competing to win. The Americans were involved, so there was a big check. I mean this literally: The check was physically big to show to the cameras. In that kind of situation, I am usually looking at my watch to see when it will be over.

THAT GREAT EVENT did not repeat itself until twenty-one years later. By the end of 1982, everything we had built seemed to be falling apart. It appeared that my young offspring would be a flash in the pan and foreshadow some sad, sorry trajectory of other inhabitants on the Island of Abandoned Toys—the Pet Rocks, the Beanie Babies, the Tickle Me Elmos. All had their moments in the sun and became neglected in clearance bins. Of course, it bothered me. As nothing prepared me for the success, so, too, did nothing prepare me for its failure. But the failure was not mine.

Looking back, I can see that the failure had nothing to do with the Cube itself, but everything to do with the way its global business was managed, which never kept up with the demands of the market. Therefore, counterfeits and copycats were able to flourish. As dismayed as I was to see the shoddy construction, the flawed designs, the terrible cheap materials, part of me had to admire the sheer audacity of these businesses, especially those from China, as they stole my product and got it out to market. And, naturally, these versions were not only cheap in construction but also in price. As these copies proliferated, the more expensive "official" version languished in piles of unsold merchandise.

The market got saturated. Cubes had entered the homes of most families who were likely to want one, or more than one, for every member of the family, kids and adults alike, so the Cube simply needed time for a new generation to appear. A craze is like a fever: It can't sustain itself constantly and needs to quiet down. But in the middle of it, no one thinks strategically about what might happen five or ten years into the future—the immediacy of the demands require so much attention that nothing else is visible.

All of this I can see now, with hindsight. When it was all happening, the collapse of the market was simply confusing.

I had slowly earned some money from royalties. Having surplus money was highly unusual in Hungary in 1982. More unusual at the time was that I decided to put my nest egg to work. I developed Rubik Studio as a small cooperative for technical development and design. I looked around for a proper building that would work for the studio, and ended up buying some old property from a church. I also made a charitable foundation for inventors and designers. There were two programs. The one for inventors targeted those who had some creative ideas but lacked a workshop or background that would help them find partners who might invest in their inventions. The other one was to help students, young designers, to go abroad and discover the world of design.

During the craze, I had been obliged to follow the Cube. So I had gone on a kind of sabbatical without pay from the university. Now I returned and began to lecture again.

I also worked on several new projects and revived some slumbering ideas that I had created before the Cube started to dominate my life.

THE SNAKE WAS A PRODUCT I made before the Cube.

There was an art exhibition, and I created a red-and-blue snake from wood. It was a special kind of constructional toy. Not in the sense that you put together elements in a different way in space—LEGO is the perfect example of that; there are elements, they fit together and can form a construction. The elements of the Snake are all connected, and you can only turn the neighboring elements relative to each other. And each element is one cube cut in half diagonally. Because of the regularity of the elements, and the regularity of the connections and the movements, what is created at the end keeps the geometrical character.

In a sense it is close to the Cube: It is geometrical and twisty. But there is no formal solution; you can make patterns of forms and create a huge number of combinations. The Snake has huge varieties of forms. You can have a dog, a cat, a staircase, a ball, and of course a snake; your fantasy is the only limit. I have never seen a mathematical analysis of how many potential configurations are possible, but the Snake does have some kind of sophistication. People who found the Cube maddening can spend hours playing with the Snake.

The classic version contains twenty-four elements, two colors, half and half. As there are twenty-four hours in a day, half is dark and half is light. It is a challenge, but not a puzzle. A puzzle is predicated on a precise question that requires a precise answer. In the Snake, there is no question but many different answers; whereas, with the Cube, there is a single answer but an enormous number of ways to arrive at it.

In the BCE (Before the Cube Era), the Snake was not a commercially viable product. But after the Cube was on the market, the company that was manufacturing the Cube in Hungary was ready to make the (patented) Snake. Ideal Toy didn't want to sell the Snake. Until then the Snake sold well in Hungary and other neighboring countries, but not in the West. Now it is part of a product range of the Rubik Brand known as Rubik's Twist; the fakes are selling as Magic Snake, with many varieties of colors and sizes.

I ALSO MADE A FLAT PUZZLE known as Rubik's Tangle.

All the pieces are identical in form, but not in coloring, which on each piece is unique. And the set of twenty-four pieces shows all the possibilities that can be done with these kinds of patterns. There are images that look like four ropes in four colors on cards made out of plastic. The sides need to all be connected by the ropes; two connect neighboring sides, the other two the opposite ones. We made four different sets of twenty-five, adding one duplicate, and if you can find a solution individually, then you can find a solution for all of them together, filling a 10x10x10 square.

With the basic set, you can cover the surface of a 2x2x2 cube; the same color ropes are connected, forming four endless loops each, and it has only one unique solution.

I view this period of the mid-1980s not as one of failure but of hibernation, with important lessons to be learned.

It became clear to me that if something is truly worthwhile, there may be an early period of success, and there may be a period when nothing much happens. But it is what comes after the success and after the failure that really matters. The resurgence is an opportunity to assess what had happened, to figure out if there had been any mistakes (because isn't there always a mistake that can only be seen in looking back?).

We need the patience and persistence to give our creation room to hibernate and revive, for there to be new potential for it to be discovered again, a fresh chance, and for the "zeitgeist" to shift. Time is not something we have to pick a fight with.

Time is something we must put to use, the same way as we breathe air. It is part of us, not our enemy.

FAILURE IS NEVER PLEASANT, of course, but for me it is an essential component involved in any effort of learning by doing, and as such, it is actually a positive thing intellectually, even if painful emotionally.

There is nothing more instructive in life than failure, and in many ways even more so than success. One must be brave enough to make mistakes, because without making mistakes, it is impossible to do everything really well. You can't do anything perfectly the first time. In my view, the key is to view failure as part of the creative adventure and to seek to understand its components. This becomes much simpler if the process is viewed incrementally, which means not setting our sights in a single-minded way on some specific goal, but shifting our focus and curiosity on each step of the way, each element of our progress.

In most cases, it is impossible to isolate the single variable that likely caused the failure. There are many components in the creation of anything—whether an interpersonal relationship (a love affair, a friendship, a marriage) or an invention. In hindsight, it is actually much easier to determine why something failed than to realize how or why something became a success. One small reason is that no matter how scientifically the components of success can seem to be calculated, a very essential ingredient is good fortune. I know that there is an American saying that one creates one's own luck, but—as a very lucky man—I don't think that is always the case.

If you discover why a failure occurred, then this wisdom

can contribute to fixing what went wrong—though of course this doesn't act as an insurance policy to ensure that you don't have a flop in the future.

WHEN I TURNED FORTY in 1984, my daughter, Anna, was six; my son, Ernő Jr., was three; and my Cube, having survived a tempestuous childhood, was already ten. I have never been preoccupied with the various symbolic aspects of aging, the famous midlife crises. But I must be honest: My transition to middle age was not without some challenges, some important moments of reckoning. Certainly, the small measure of fame and fortune that I achieved in my thirties was unlikely to be replicated. But it was only when I had gone through the storm and the calm that followed that I felt able to process what success really meant. Was my success rooted in the great commercial popularity the Cube had achieved, or did it occur long before then, when I had discovered it and realized that I could solve it?

I suppose a bit of both. Success is a strange phenomenon.

It seems everyone hopes to achieve it, but what does it mean? The commonly accepted definitions—having a high position, being part of some elite order, being rich, having the admiration of others—may constitute what is generally understood to be success. But for me it does not even get close to the most important aspect or significance of the term. Other definitions come closer to the mark: "Success is the achievement of something you have been trying to do," which means that something works in a satisfactory way or has the intended result. This resonates more for me because it suggests a relationship between an individual and his work, rather than how an individual is regarded in society, or in any public sphere, which involves many other factors that have nothing to do with achieving something.

SINCE I AM FASCINATED by contradictions, I enjoy the fact that the Cube is a healthy microcosm of both success and failure.

As a product, of course, it was and is a great success. But let's distance ourselves from conventional measures of success, which tend to be those that are easiest to count. What if the Cube had never become a worldwide sensation? What if it had a small measure of success in Hungary but then disappeared? I would have continued teaching. I would have continued to design. I would have traveled less. And I would have continued to consider the Cube to be an accomplishment.

For me, the very creation of it, in itself, was also a success. Each phase of the process contained mini triumphs and moments of exuberance. When I was able to make all the cubies hold together, for instance, or the moment I discovered how to make it move in so many different directions, or the moment I watched the colors scramble, and then the moment when, after working on it for a month, I was able to bring order to the chaos that I had created by suddenly solving it.

It is a special experience of success available to anyone, that is repeated millions and millions of times.

In *Popular Science*, a writer described her triumphant feeling when she first succeeded in solving the Cube. "Back in the office," she wrote, "I can't say for sure what I've learned. If I kept practicing, would I just get really good at Rubik's cubes? Or would I find that other, related skills grew, as atrophied parts of my brain bulked up, and all manner of spatial solutions made themselves known to me? I don't know. But in that moment when all the pieces locked into place, when I'd restored order to this scrambled shape, my mind was clear for a moment. Instead of the regular worries of the day, I was filled with a sense of endless potential."

I know exactly what she means. My cathartic experience of success occurred almost fifty years ago, when I was a young man in my room, earning close to nothing (about one hundred dollars a month), trying to figure out what I had just created.

CONVENTIONALLY, THOUGH, in today's world, success is quantified by something else: money.

Money is enmeshed in our everyday lives; it appears and disappears in a thousand disguises. Our relation to it is as complicated as our relation to other people, and maybe that is not surprising, since it has become the symbol of value and power in our world. Our emotions are contradictory toward money. At the same time we can love and hate it, respect and despise it, fight for it and forsake it. Different nations speak of the many possible relationships to money when they speak: The French "gain" money (gagner), the English "earn" it, the Americans "make" it, the Russian "works" for it (работать), and in my mother tongue, we "look for" it (keres).

Our attitude toward money is characteristic of each of us, and so, too, is the extent to which we are keen to find out the financial situation of others. If one accepts that money is the best way for measuring the value of things, it can easily become a compass for finding one's way in the world. This is rather convenient. But it is dangerous and misleading. There is also the delusion that since money is so quantifiable, it must tell us something unambiguous. Nothing could be further from the truth.

ONE OF THE VERY FEW BENEFITS of growing up within the economic system of state socialism was an overall disregard for financial gain. Outside of the Party apparatus, people were given little incentive to work very hard. Those who did go the extra mile could still not live much better than those who just took it easy. And yet, some people, like my father, worked very hard indeed. What pushed them? And what pushes anyone to immerse in a very demanding and often frustrating challenge like the Cube that offers no benefit beyond its own solution? I have never stopped wondering about the mystery of human motivation. When survival and even well-being are safely established, why do people still tirelessly strive to go further and further?

The Cube's astonishing success had made me comfortably well-off before I turned forty. I never was truly wealthy, but I always had more than I needed. (And I never really needed much.) My only expensive "hobby" was to build our family homes exactly the way we liked them without compromises. Otherwise, I have always enjoyed home cooking much more than posh fine dining restaurants; preferred our local Lake Balaton to exotic journeys; and been happy driving my eighteen-year-old Ford Galaxy until very recently, when I downscaled to an electric Golf. As such, making (more) money was no use for me as an incentive to do anything much.

Another reliable motivation from our evolutionary history is the urge to achieve comparative advantage over our peers. One can aspire to be richer, smarter, stronger, or more beautiful than one's neighbors or colleagues, or even more than the remote people known only from their television appearances. Alas, this competitive spirit did nothing for me either. I have always been a slightly reclusive person, an introvert, and I could never truly identify with any specific group populated by people with whom I would compete. I have made some money in the toy industry, but nobody would think of me as a toy professional. I didn't have a "career path" with highs and lows—I didn't even have a clearly defined occupation for any sustained period of time.

Beyond fame, fortune, or relative success, there are noble incentives that motivate some exceptional people. There are true heroes who relentlessly follow their calling to fight climate change or eliminate poverty, cure dangerous diseases or increase animal welfare. Unfortunately, I must also admit that while I deeply respect them, I am not among this select few. I have always tried to help those who sought my advice and I have contributed to ambitious educational projects, but it was up to others to create those institutions, run those foundations, or foster those talents.

So what is it that drives people like me: introverted, peaceful amateurs without either well-defined long-term goals to achieve or short-term needs to fulfill?

PSYCHOPATHS ASIDE, there's certainly a sense of responsibility that makes you feel obliged to assist and participate in others' plans and priorities, especially if they are close to you emotionally or socially. For example, I have never really enjoyed the publicity stunts, formal engagements, or sales meetings I have had to do for promoting the Cube. However, many people working hard for its success relied on my support and presence and I obviously owed them that much (and then some).

I remember reading once about motivation and a concept

called *intrinsic motivation*. Some psychologists divided this kind of motivation into three aspects: motivation toward *knowledge*, toward *accomplishment*, and toward *stimulation*. It seems that when faced with a Cube, all three of these traits are present.

Deep down, however, my motivation has always been anchored in childhood curiosity and a quest for understanding *how things work and why*.

All children are wonderfully motivated, and there's really nothing like playful curiosity for learning. In their natural habitat, lions have little to fear. Sure enough, adult males typically spend their days dozing in the shade, using precious energy only if there's some food available (brought home by the female hunter) or they are driven to mate or fight for status with competing males. But just look at the lion cub that still has to learn everything in order to rule his animal kingdom when he is grown! The cubs play tirelessly with no regard to heat or hunger. This is how they acquire both the knowledge and the skills that will ensure their survival when they grow up. Humans, fortunately, do not have to save all their energy for later. They can afford the luxury of remaining playful and curious all their lives.

Still, very few are lucky enough to nurture their childhood curiosity into adulthood. *That might be because intrinsic motivation diminishes by external rewarding.* When kids start learning in order to get better grades at school instead of immersing themselves in open-ended questions, they quickly understand that it is the grades that matter, not the learning.

External rewards and punishments are astonishingly effective instruments for changing people's attitudes and even interests. Therefore, those who have the means to offer such incentives also bear an enormous responsibility.

Schools can cultivate (and discourage) children's interests the way they deem useful for society. Company owners can create incentive packages for their executives to orient their focus on what is perceived as important. And it always works. Except, if the school gets it wrong, society may lose a potentially great artist who instead becomes a mediocre accountant. Great family businesses may be driven into the ground when detached executives are given unwise incentives. It is at this great divide between external and internal motivation where it is most useful to distinguish between amateurs and professionals.

An amateur finds satisfaction in the very task, activity, or problem at hand. A professional is guided by the external rewards his success at that task may yield. If the reward is taken away or the balance between effort and incentives changes, the professional may just move on. For the amateur, nothing changes: As long as the problem excites them, they will stay with it. At that particular moment in time, that very challenge is not interchangeable for any other: That's what the amateur simply wants to pursue.

Of course, an amateur's interest can also be surpassed by external factors.

If a talented but penniless painter works on a groundbreaking conceptual work of art but is offered a fortune for an advertising campaign instead, he might well be lured away. This is exactly how incentives can corrupt and why they need to be applied with the greatest care—even while appreciating real life demands.

This vast power of incentives underpins most organizations as well as society at large. To be fair, this is what makes professionals more reliable than amateurs to meet deadlines and respect the wishes of their clients. Professionals can deliver that famous "bang for your buck." Amateurs are more of a gamble: They might come up with something extraordinary, but they may just as well follow a path that is of interest only to themselves.

Steve Jobs once said, "I was worth about over a million dollars when I was 23 and over ten million dollars when I was 24, and over a hundred million dollars when I was 25 and . . . it wasn't that important—because I never did it for the money."

In Apple, Jobs built the world's most valuable company, and he was arguably the single most influential designer and tech evangelist of our time. He was a billionaire several times over. Even so, I still believe that he was not fundamentally interested in money. He had a very clear vision and he pursued his world-changing dream. In order to achieve that kind of outstanding impact, he needed that power and those funds at his disposal—but only as an artifact, not as an end goal.

So was Steve Jobs an amateur? This is where the distinction crumbles: His vision was not interchangeable for any other, and he surely would not have changed his interest for anything.

He was simply the quintessential professional with intrinsic motivation!

MY APPROACH TO MONEY has always been to ignore it. My sense of money, if I have any, is a feeling that money is kind of harmful, because it creates work, you have to make a special effort if you have it, and you need to look after it and handle it. And that is not something that inspires me.

I often think of an old story about Northern and Southern Italians. The Northern Italian, very professional and businesslike, goes on holiday to the south and sees a Southern Italian sitting on the seashore and trying to catch fish. And the Northern Italian arrives in a nice car and starts to have a conversation. The Northern Italian talks about himself and how successful he is, and now he can enjoy all the fruits of his labor and come to the South and catch fish. And the Southern Italian guy shrugs and says, "I've just been doing this my whole life, and I'm here too—nothing special about that." The Northern Italian had worked hard to achieve the goal of what the Southern Italian took for granted.

In the end, there are many measures of success, of which, apparently, money is the easiest to quantify, but for me, it reveals the least of anything important.

Being an inventor can never be a conventional job. The very nature of inventions is that they are original, which also means something extraordinary and unpredictable. Earning pay is what makes a job a job. But one cannot assign a value to what constitutes the essence of inventions: the original concept. Not because it is more important than a job, but because it is different.

Even in the early years, when the very first signs of success appeared, I was taken by surprise. Already in 1979, my income from the Cube was greater than any salary I had ever received. Intellectually, I realized that my bank account had grown, but it took much longer until the implications of that growth registered emotionally. It was difficult for me to imagine that somehow this growth was connected to the value of the Cube.

People speak about value and price and assume that value is the same as price. There is some kind of relationship, but one is far from the other.

Usually, as someone's money gradually increases, they adapt to their new circumstances, acquiring and doing things in accordance with the new situation. But the change in my fortunes happened suddenly for me. There are so many awful examples of how unexpected wealth ruins some people. The lottery winners who squander it recklessly or do not find their way in new surroundings and ultimately fail in remaining true to themselves. Or we see artists, creative professionals whose very creativity is affected if success comes at a time when they are not ready for it—it can kill their spirit. There are so many traps, some visible and many invisible.

For some reason, I have managed to avoid them. That is probably because, as I said at the beginning, my approach to money has always been to ignore it. Money isn't really important to me. When I didn't have money, I had no problem because I had what I wanted. For me, one of the keys to happiness is to not need more than you are capable of. I suppose that someone could say that this is the luxury of having it. But the fact is, even when I was living on my university salary, I felt the same way. In the end, I will always be my father's son, dragging rocks from one shore to another, just to make something useful.

5

Success is not the key to happiness.
Happiness is the key to success. If you love what you
are doing, you will be successful.

—ALBERT SCHWEITZER

RUBIK IS NOT A COMMON NAME in Hungary, nor else-where. It took a while for me to get accustomed to seeing it so often in so many different contexts. Soon there appeared Rubik furniture, and all sorts of products that had nothing to do with the Cube but had my name on them. That is the nature of people and communication and marketing. If they call furniture "Rubik furniture," they are using my name as an adjective and not as an individual's name. And that adjective conveys . . . what? That its design is sleek? That it has bright colors? That it is connected to the Cube and all that it means? That if you buy the furniture, you become a member of an exclusive Rubik club?

Of course, none of this is true.

Anyone can have fame; even a serial killer has fame. But what does fame mean? That there are a certain number of people who know of you or have heard about you or know what you did or didn't do?

A paradox. The whole world seemed to know my name, but my name was no longer linked to me as a person. My name had become public property.

As the years wore on, this strange phenomenon began to intrigue me as a different kind of puzzle. After a name becomes a brand, what happens to the human being with whom it is connected? I remember once reading an interview with Calvin Klein in which he said that he met many people who did not believe that he actually existed as a human being. His brand name had become "independent," completely disconnected from him!

The Cube entered a small club of inventions each bearing their creator's name, and over time, my name, too, became so wedded to my invention that the personal connection has been all but lost. Included in this pantheon is "diesel" (Rudolf Diesel, 1858–1913), "macadam" (John Loudon McAdam, 1756–1836), "thonet" (Michael Thonet, 1796–1871), "gillette" (King C. Gillette, 1855–1932), and "biro" (my fellow Hungarian László József Bíró, 1899–1985). It is strange to realize that many people who own a Cube have no idea that there really is a man named Rubik—much less that he is still alive.

One thing that happens with fame is that many people think they know you. They feel as if you are almost a friend or a neighbor; they imagine personal details about your life that may or may not be true. There were times when people thought I was the richest man in Hungary. Then there were other times when people thought that I had no money left. That I might have been prosperous once, but that was over because of unscrupulous people around me. The rumors begin to get traction. Shards of my biography were grabbed and displayed as if they were facts. But facts were rarely involved. If the Cube is secure in its public persona—demanding, limber, compelling, addictive—mine is muddled beyond all recognition.

I AM ALWAYS BAFFLED when people seem to be eager for fame. Fame is not what my type of person would desire. Like the great actress Greta Garbo, who hid after she became famous, I, too, like to stay in the shadows. Naturally, there are people who clearly enjoy being in the spotlight. They love to be with thousands of fans and competing with other famous people for "likes" and "followers." Some people actually seek power through fame.

Once, in the eighties, when I was in Japan, a rather large crowd awaited me. Mothers brought their kids to shake my hand with the strange belief that my "power" would be transferred to them. I visited several large Japanese cities and would appear in gigantic department stores. From the first floor all the way up to the top floor where I was sitting, people waited for a long, long time to come and shake my hand. I cannot estimate how many, but perhaps thousands of people stood patiently in this long line and one after another, they would approach me and extend their hands. I felt a bit as if I had been put in a cage with a "Don't Feed the Animals" sign on it.

This was what life was like for me in the early 1980s, and I found it a bewildering experience. I was just a teacher of architecture and design in a country behind the Iron Curtain. This was not yet a time of "globalization" as we know it now, so I was faced with trying not only to comprehend and grasp the success but also to assimilate all those bizarre, peculiar, exotic places that success took me to. As the craze took hold, I became a kind of commodity, required to visit toy fairs and make media appearances. I felt oddly dissociated from it all. I may have been physically present, but my relationship with all that was going on was closer to being an observer. That was the way for me to survive that type of life more or less unharmed.

I have strived to minimize the negative and maximize the positive elements, especially when the tidal wave of fame engulfed me early on. Wherever I happened to be, I reminded myself that it was a transitory situation. One of my favorite Hungarian sayings is, "All miracles last three days." This miracle has lasted not just for three days, or a few years, but many decades already. That continuity, year after year, especially after the turn of the century, is another series of wonders.

I NEVER WANTED TO become an inventor. I never "wanted" to become anything, really. I had no vision about my personal future; my present occupied me.

It never even occurred to me that being an inventor actually *was* a profession. I was interested in math but knew I would never become a mathematician. I liked mechanics, creating instruments, taking apart and putting things together. But I knew that I didn't want to become a mechanical engineer. I didn't want to have any single profession; I wanted to have all of them. And probably that desire is what led me to architecture.

I am not widely known for anything that I have designed or built—aside from the obvious. There are not hundreds of buildings in my collection of works, because I never was a working architect in a firm; I was always a teacher, and that was my main occupation for more than twenty years. Of course, I was involved in a few buildings, but they aren't that interesting. The ones that are most interesting are the buildings that I designed for myself. The homes that I have designed are all expressions of who I am, or of who I was when I built them.

As an architect and designer, I worked with teams and with systems. Members of a team need to understand each other, and they need to be able to communicate with each other. There must be a structure in which one person is responsible for what team members are doing individually and what they are doing for the team as a whole. When it all went well, working with all the components, all the other specialists, coordinating everything that went into creating a building felt like being a conductor of a great orchestra. The conductor needs to know how to play every instrument, even though, with a few notable exceptions, he may never do so in public. The same is true for an architect and a designer, even though the sheer number and variety of all the professions involved in making a building a reality is much greater; it is a process that metamorphoses day by day. I have always loved the collaboration, watching the previously disconnected elements come together one after the other. My ambition was not to create grand public spaces but private homes—spaces that needed to respond to the daily lives, routines, and intimate moments of their inhabitants.

MANY PRINCIPLES OF ARCHITECTURE AND DESIGN seem to me to have much broader applications for managing the problems that life throws in our direction. To solve most problems, in life and in architecture, one needs to know how to cooperate. The plan for a building is not just personal but involves exchanging information, whether documents or drawings. There is the old story of a group of blind people surrounding an elephant and attempting to describe it by touching individual parts. One person who tries to find the perimeter of an enormous leg imagines she is faced with a rare kind of tree. The person who has the trunk imagines that it is a snake. It is impossible to see the whole animal by focusing on the parts. One must see the connections among them. Without seeing the underlying system, it is almost impossible to understand what something really is.

If you are walking inside a building, you will never see it as a whole. In order to do that, you must fly over it. You can have an image of the structure, and for me that means more than a visual impression of its appearance. You need the contents as well. I will have built many homes in my mind by the time the first blueprint is drafted.

When I design a house for a family, the family becomes the most important part of the team—even if they have no knowledge about architecture. I need to learn about them and begin to visualize who they are and how they inhabit their space. I need to reach out to them to better understand their views, just as they need to understand my ideas as a designer. If this cooperation works, the result will be good. I feel connected to them, as if I were one of them.

I have spent a lot of time thinking about our built environment and how an individual is affected by his or her surroundings, both the personal and the larger one, both of which can create restrictions and offer opportunities. If you consider the great possibilities for culture and comfort in a city, you must also confront the limitations. Then, within the larger environment are all the opportunities to create smaller, more personal ones. Architecture also influences and creates environments, usually within already existing ones. If your task is to design a home for a family, the optimal case is if they have not even chosen the plot of land. If they haven't, there is a wonderful chance to be part of the entire process. But architects can engage with the project at any phase of it. Naturally, the later you join, the less you are able to accomplish, until, finally, you may only be able to take a look at the interior design, and decide about the lighting, the colors of the walls, the best carpeting and furnishing.

The built space is physical, of course, the space of a house. But, as it becomes a home, it is also very emotional. Aside from questions about materials and style—Wood or stucco? Bricks or stone? Modern or traditional?—there are other, subtler but equally important ones. When one looks out of a window, what does one see? The surroundings are usually not homogenous, but rich and changing. Where is the sun? What is the orientation of the building and what does it mean for the morning and evening light? What direction does the wind generally come from, and how do the shadows fall? What does it mean to have an open, yet protected, space? And what are the best proportions for the various rooms?

Often, people find themselves living in spaces in which they feel uncomfortable, for reasons they can't really describe. Humans are complex, with so many individual tastes and preferences and emotions. But the heart of the problem is the recurring theme of space blindness: We are not really trained to see the world around us and how we fit into it. Understanding our relationship to space, how we should look around to actually see the surroundings, never appears in a typical school curriculum. As a result, unless one is trained in architecture or design, or in fields like art or choreography, we very seldom understand how to handle our intimate and our more public spaces. We don't really know what we are looking at, or how to look at what is all around us.

Indeed: What does it mean to feel at home?

THE PLACES IN WHICH I HAVE LIVED symbolize very distinct times in my life and the life of my family. My first home was a sprawling apartment on Szent István körút in Budapest. I lived there with my parents and then with my mother for nearly twenty-five years, from when I was five until I was thirty. It was a traditional Budapest apartment building, from the early twentieth century, with an internal courtyard surrounded by several stories of flats. The more expensive and prestigious apartments faced the street, while the more modest ones faced the courtyard. There was a main staircase, an elevator, and a smaller staircase on the other side of the courtyard for the servants. The flat in the corner back was ours.

This was once a much larger apartment, but it then got cut up because of the political changes after the Second World War. Originally, the flat had a room for the maid with a separate entrance, next to the large kitchen, then an entrance hall, a large parlor, and a living room. There was space for parents, and the children had separate rooms. The flat was on the first floor, so a separate entrance opened into the garden, which had large acacia trees facing south. But then, part of the flat was cut off to house two new families who got their own separate entrances. The original hall was my room for a long time. After my parents' divorce and after my sister moved out, I was already at the university and had the corner room, which I immediately made my own. I made my bed and painted the walls and put a blue carpet on the floor.

My father had moved to his own traditional house, with a

high sloped roof and a retaining wall between the house and the street. He had decided that he wanted to build an addition and had already begun the project. After I married and achieved some early success, I planned to move there with my wife. The Cube was ready, but it was still only a lowercase cube. The addition to the house was the work of an amateur, and my father was not able to finance it. So I told him I would use my expertise to both finish it and finance it. I then used the potential of the roof to create a space that was originally a one-bedroom flat with a nice large living room and a separate bedroom. After my first daughter was born, I divided one part of the living room into her bedroom.

BUT THEN MY LIFE CHANGED as the Cube took off.

I also divorced and remarried. I bought a house that had been designed by a famous architect, created in the wood-and-stucco style of traditional Transylvanian architecture. During the war, part of the building had shifted and there was a crack in the foundation. But it had a very beautiful structure with a high roof. When I made the new foundation, I also built a swimming pool in the basement. I took advantage of the high roof to create a second floor for the children. It was the first building that was really all my own. I tried to keep the original atmosphere while upgrading the windows and doors. I even designed the furniture. And we created a beautiful garden. We lived there for fifteen years, until the turn of the twenty-first century. By then our youngest child was almost ready for college but still living at home. I suppose we should have downsized, but instead we moved to a new and larger house.

There are people who live in twenty places during their lives. And there are people who live only in one. I am both. I live in one place for what seems like a long time, and then move to the next one. This time, I wanted to build a house entirely from scratch. There was nothing on the plot of land but an old house that had been destroyed by war and time. It was neglected and abandoned, so I demolished the building but used all the old bricks to build a new house.

I wanted something new, but in the end I designed a quite traditional house, adhering to the design rules of this area, a very green suburb of Budapest, next to the protected forest. Budapest is a town divided by the Danube: Buda is hilly, green, spacious; Pest is flat, dense, citylike. In the early days, people of means who had big flats in the city would move to Buda for the summer. These buildings traditionally have stone walls for the ground floor, and they are very stable. I used the bricks for the structure and made a high tiled roof with an overhang that protected the unplastered brick walls from the rain. It also created some beautiful shadows in the house when it was sunny. The lot was very green and beautiful, with large old trees. And once more, we had a lush and luxurious garden; when the weather was nice, the outside became almost another room.

I didn't build a swimming pool this time but added a separate greenhouse; that was when I began collecting succulents and cactuses, and today we still have a collection of them, from the smallest to some that seem like trees.

Eventually, our youngest daughter moved out, and the house and garden started to be way too big and was too demanding to maintain. We felt it was time to go higher up and get some more sun and a wider view. A property nearby that seemed like the kind of place in which a smaller, more modern house would fit well was for sale. My wife has a very strong sense of design, so this was a house, perhaps our last one, on which we collaborated very closely. It is less traditional and very minimalist. We used materials like granite and stainless steel, which need little care and last a long time. The house is black with some white, which is a bit strange, and vertical—five stories—so, in order to avoid the dangers of steps as we grow older, we have an elevator. It is a house for people who have no children at home. But since we have already six grandchildren, there is a recreation area in the basement for them where they can play.

I believe that someone's home can express more about the people who live there than can be found in a conversation with them. I always feel as if I learn so much about people by looking at their home, which is very different from looking at just a building. As I look back on my four homes, each one says something about a certain period in my life. If I close my eyes, I can wander through each and every one of these and see the shadows on the staircases, the views through the windows; remember some critical moments in construction and design; and see my children when they were young, sitting around the dinner table.

TO MAKE A COMEBACK, something, or someone, first needs to go away. There is always so much talk about the Cube's "comeback," but in my world, the Cube never went anywhere. Despite the *New York Times* obituary of the Cube, I never felt that way. I never looked at the world as a commercial arena, so the Cube's success was not measured in dollars. During the eighties, I was not a director of the drama. I was a mere actor, and not the star. Being part of the repertory, one has no idea how many people there are in the audience or how many tickets are sold. So I didn't have that kind of external perspective either.

The fact is, the Cube was still there, even without the interviews, the exorbitant sales figures, the massive media attention.

The noise was less. The inventories were filled with Cubes. We are speaking about a three-year period. It was declared dead in 1982. As it turned out, it was not dead but only sleeping. By 1986, there was a resurrection.

At that time, I was still young. I was not at the end of something; it was the beginning of life. I had no interest in being a has-been, because I wasn't. So I created other puzzles. I managed a foundation for inventors. I lived a fulfilled and happy life during that period, and meanwhile, the Cube was plotting his next act. The early signs were in 1986, when we began working with a new distributor. Curiously, the stock that had been languishing in stores disappeared because people began to buy it, and so new orders appeared. But the enduring success began in the mid-1990s, and by the turn of the century the Cube had become stronger than ever. That doesn't mean repeating the craze. It was not a fever nor an expression of nostalgia; this growth was stable, healthy, and continuous. A new generation, some of whom hadn't even been born in the early eighties, discovered it when they were teenagers in the nineties as something that was both familiar and novel.

EVER SINCE THE CUBE ENTERED MY LIFE in Budapest, in 1974—but especially as his celebrity status increased—I have felt a bit like the old wood-carver Geppetto, watching his creation come to life, full of mischief and a sense of adventure. Much like Pinocchio, my Cube became autonomous. Not only has it become independent, it has revealed himself to be in many ways my opposite. The Cube loves attention; I don't. He is eager to interact with everyone; I sometimes find this a bit difficult. He's quite ambitious; I am less so. I may sound odd describing an inanimate object this way, but I believe that every cuber, no matter how new to the process, develops a kind of relationship of their own with the Cube. At the peak of the first craze, a book entitled *Not Another Cube Book* was published and one chapter was titled "We Hate the Cube." For me, this was the strongest possible evidence that people not only enjoy the Cube or simply like it, but they love it.

How is the Cube capable of evoking such strong emotions and attachments? There are a few factors that help create these almost familiar relationships. One key is its ability to move. The connection between life and motion is so strong that even though we are aware that nonorganic things are also capable of movement, we tend to describe them as if they were living. Giving something a name is an expression of a connection; indeed, it is the way we connect. In the natural world, the freer the motion, the stronger the tendency to anthropomorphize: a bird is more alive than a tree, although we often apply human terms to trees as well, with their roots "gripping the soil." The colors and movement, the warmth it picks up from the hands that hold it, all generate the Cube's dynamic sense. And with that dynamic sense, as with all beings that have a strong dynamic sense, I suppose, comes an extremely strong will. The Cube is stubborn; he does not surrender. We conquer him only by learning to speak his language. And increasingly, many people became fascinated by learning the language.

AT FIRST, I BELIEVED THAT THE CUBE was nothing more than the expression of an idea, and nothing more than itself. It was both the embodiment of a thought and the expression of a way of thinking. But as the Cube has spread throughout the world, it can be harder to peel off all the layers of cultural, commercial, or artistic impact that have appeared during its journey of more than four decades.

During the craze, I began collecting as much as I could about every mention of the Cube internationally. It is something that I continue to do today, but I know that even with an overwhelming amount of material, my record is far from complete. I have enormous files containing all sorts of references: magazine covers, appearances in books and movies, in art, in architecture, in design, cubing championships, records, the fastest time solving it with various appendages and in various states. There are competitions and clubs, paintings and sculptures, graffiti and sidewalk art, rap music and classical quartets and symphonies.

At the 2009 Burning Man music festival, in Nevada's Black Rock Desert, I'm told people could sit beneath the giant Groovik's Cube. It is a fully playable, thirty-five-foot-high sculpture inspired by my puzzle. With about 15,000 elements, it took less than five months for Mike Tyka, Barry Brumitt, and a team of artists to build. My collection has expanded to include many thousands of articles, newspaper clippings, typed manuscripts and formal publications, bank account records, interviews, reports, letters, contracts, programs, airplane boarding cards from my travels, agendas, foreign currency, posters, program notes, patent cartoons, certificates, and crazy, funny, and ponderous objects that are difficult to classify.

The Guinness World Records has an astonishing number of entries where the Cube plays a starring role. Naturally, there are many speedcubing records. But there are a whole host of records for people with certain talents—holding their breath or skydiving or being clever with their feet or juggling or doing things with their eyes closed or standing on their heads—all while cubing. For these people, the Cube becomes a kind of megaphone, a way to emphasize their personal skills or hobbies or idiosyncrasies.

The Cube has also become an artistic muse. It appears in the Museum of Modern Art, both as part of the collection, as something like an original piece of art, and also in the gift shop. Even though I am responsible for the colors and the design, as a piece of art, it exists in its own domain, somehow independent of me. Increasingly, it seems as if the Cube had been waiting to be discovered, and I was the fortunate person who stumbled upon it.

In that realm, the Cube is a work of art because it has a kind of perfection. The great French writer and aviator Antoine de Saint-Exupéry described this state as being "achieved, not when there is nothing more to add, but when there is nothing left to take away."

But of more interest to me than the MoMA exhibit are all the paintings, sculptures, drawings, photographs, multimedia objects, and murals from artists all over the world referring to the Cube or including it as part of the medium. Through it, they discover ways of expressing something original and exciting. An internationally known French artist, who only goes by the name Invader and wears a mask so he cannot be seen, has built his entire career over two decades on using the Cube in his paintings and mosaics. He began as a street artist, which is why he called himself an invader, and because he has also "invaded" the space of major galleries and museums. He has created over 3,500 works of art, and in 2018 there was a big retrospective of his work in Los Angeles. No one knows his real name, but when we met once in Budapest, he removed his mask, revealing a very appealing-looking young man, so we could talk.

There are many others who have created Cube art all over the world. I have seen paintings where the Cube is the subject, nothing else. There are paintings where people are playing with the Cube. There are paintings where the Cube is the head on a human body, or where there is a Cube body with a human head. A former Italian speedcuber and artist named Giovanni Contardi creates mosaics of his favorite artists, celebrities, and TV and movie characters out of thousands of Cubes. Some artists, and Invader is one of them, replicate classic works of art, like *The Mona Lisa* or *American Gothic*, into mosaics made out of Cubes. In Toronto, there is a studio known as Cube Works in which artists gather, spin Cubes, and create massive murals. One of them was a version of Leonardo da Vinci's *The Last Supper*, which used over 4,000 cubes. Large Cube sculptures appear in public spaces, in street art, in museums. In University of Bristol there is a sculpture of a monkey playing with the Cube. There was a Chinese exhibition in which the artist used the Cube as a symbol of diversity by having a whole pile of Cubes in the middle of a gallery, with symbols of various religions encircling the exhibit, as a way to show the diversity of our culture. I have even seen a picture of the Cube on a gravestone.

The visual arts are not the only places that have been inspired by the Cube. The Cube has been used in lyrics and album covers; and even inspired experimental music. Popular songs have used it to speak about some feeling about life.

The advertising industry has found the Cube irresistible, and it has been used to sell almost anything: computers, plastics, thrillers, banks, cigars—even condensed chicken soup. I have had no control, much less any profit, from all those who have appropriated the image. I often say that the Cube and I have had the pleasure of making many people very wealthy.

Once, when I was in London, a television commercial advertising soup had the tag line:

"What came first? The chicken or the Cube?"

THERE APPEARS TO BE NO LIMIT TO CARTOONISTS' ingenuity and fantasy when it comes to depicting the Cube. And once more, there are no national distinctions or differences in the nature of the humor. One from the States showed an elderly man perched on a high-floor window edge ready to jump with an unsolved Cube in his hand, his wife appealing to him desperately, "Darling, it is just a toy!" In the newspaper *Frankfurter Allgemeine Zeitung*, we see a pair of legs dangling in the air, and below him on the floor lies a scrambled Cube (no wife intervened to prevent his fate).

Most cartoons are not so macabre but almost sweet. There's one of a young girl who wears a T-shirt that says: "Going Through Cuberty."

And I cannot begin to count all the political cartoons in which Cubes stand for the unsolved problems of politicians. On covers from newsmagazines all over the world, this small puzzle appears to signify the chaotic state of our planet. At the same time—with its colors, form, the square pattern—it has become an almost reassuring presence, one that draws the eye.

However, the most overwhelming cultural pedestal for the Cube has been in movies. Edgy art films, sophisticated documentaries, and Hollywood blockbusters all have featured the Cube, sometimes in cameo appearances and sometimes in very central roles.

Early in the 2008 romantic-horror film *Let the Right One In*, we see the young protagonist, Oskar, sitting alone on a frozen winter night, playing with the Cube. Eli, a young girl, or so she appears, approaches. When Eli shows interest in the Cube, Oskar gives it to her with the proviso that she return it to him within two days. Much to his amazement, Eli—a vampire, as we will learn—returns it to him solved, kindling their blood-soaked love story and setting the events of the film in motion.

As it does so often, the Cube communicates on more than one level here. It tells us that we are in the eighties, a decade so closely associated with the craze that you are unlikely to get through an exhibit or documentary about that period without spotting a Cube. It also reveals Eli's superior (in this case, inhuman) intelligence. And it expresses what is to me one of its most extraordinary qualities: the ability to prove, again and again, that when you get down to the essentials, *we are all more alike than different*.

The Pursuit of Happyness, a Hollywood blockbuster made in the same year, featured a real superstar with the Cube. The main character, played by Will Smith, has a fateful encounter in a cab with an influential corporate executive. As the businessman watches Smith solving the Cube, he is impressed and later helps Smith get an internship—just when he needs it the most. The Cube's metaphoric reference is unmistakable and immediate: Whoever can solve it must be very smart indeed! (Will Smith, as it happens, became a very keen and very able cuber in real life as well.)

More recently, the Cube had a cameo appearance in Oliver Stone's *Snowden* in 2016. The movie portrayed Snowden as smuggling out classified information through a tile in his Rubik's Cube. But did it really happen that way? As it turns out, it did. Back then, Snowden himself hinted that there may be some truth behind the portrayal. He did use a Rubik's Cube to identify himself when he first met with Glenn

Greenwald. The movie shows him downloading files onto an SD card and then hiding it inside a Rubik's Cube. He then avoids running the Cube through security by tossing it to a security guard, who plays with it while Snowden passes through. Then he's free.

Of course, the Cube's film roles are endless, ranging from the metaphysical, as in Spielberg's *Ready Player One*, to the secretive, as in *Duplicity*, where Julia Roberts's and Clive Owen's characters recognize each other by their Rubik's keychains, etc. As for television, it is an even more natural space for the Cube. I particularly enjoyed the role it was given in the sitcom *The Big Bang Theory*.

IT IS ALWAYS FASCINATING when a consumer product acquires widespread cultural significance.

The least surprising aspect of this phenomenon is when vast marketing departments on practically unlimited budgets "sell a story" made up for more efficient sales. Some of these are very clever and astonishingly efficient. Think of L'Oréal's legendary "Because you're worth it" campaign.

More to the point is the amazing impact of smartphones on our lives, which indeed transformed our everyday habits as well as our worldview to a great extent.

And there's the Cube, which is self-contained: It didn't replace or improve on anything that existed before; its impact is all its own, and therefore all the more surprising.

But what is it that empowers a product for cultural relevance?

Of course, no one really knows, but there are some clear indicators. First of all, it must have a wide reach and achieve prevalence in a relatively short period of time. This phenomenon is much better known today with digital content— think of Angry Birds or Candy Crush suddenly on everyone's phones from pole to pole. Before the internet and universal platforms, in an age of localized markets, very few physical products could accomplish global distribution and succeed everywhere. In the toy field, LEGO was similarly widespread and very influential culturally as well.

I think it is important that the product has some *durability* and stays with the consumer for a reasonable time—although single-use products sometimes do change the course of history with some eminently essential functionality, such as certain medicines or contraceptives. Most prominently, however, consumer products of cultural value *must capture and express some rather specific meaning, unique to them* that anyone can appreciate pretty much immediately.

Historians sometimes refer to Mikhail Kalashnikov as "the other" inventor behind the Iron Curtain to become a household name globally. It is undeniable that a Kalashnikov carries a message that requires little explanation. Still, it is everyone's good fortune that many more people have had personal experience with the Cube than with that famous weapon.

Consider this: The Cube has been handled by one in every seven people of the world. And through an interesting combination of it complexity and ubiquity, the Cube became *the ultimate symbol of intelligence and problem-solving*. It carried both positive and negative emotions: the aha moments accompanied by a strong sense of accomplishment, to the head-banging frustration, impatience, and annoyance over its perceived impossibility. Culturally, the Cube is the unambiguous symbol of these concepts, even as it is an instantly recognizable touch point for the 1980s, or ingenuity with a touch of nerdiness.

Culture itself is expressed in symbols, and I was fortunate enough to experience it all on the occasion of the fortieth anniversary of my invention.

In 2014, one of the most iconic edifices of the world, the Empire State Building in New York City, was lit up in the Cube's colors in celebration.

Google's search portal featured a working 3D model of the Cube that all its users could play.

On my side of the Atlantic, the president of the European Commission cut up a huge cake in a very familiar shape to pay tribute to the Cube's origins.

One of the earlier, local anointments of the Cube was made for a previous anniversary. The Hungarian National Bank designed a square-shaped five-hundred-forint coin in celebration of the Cube. It was legal tender, real money, and I actually bought up a sizeable quantity of the coins. I still have a few hundred dollars' worth of cubic forints.

THE CUBE FREQUENTLY REPRESENTS complex geopolitical problems—think, for instance, of the *Time* magazine cover featuring a Cube painted in the US and Pakistani flags over the words "Why We're Stuck with Pakistan."

The problems in question frequently have definitive, if difficult, solutions, as when UK Prime Minister John Major used the Cube to symbolize the achievements of the Maastricht Treaty.

During the Brexit crisis, Sky News created a Brexit Rubik's Cube and asked some of its guests—including several Members of Parliament—if they could solve it.

If the Cube can signify complexity, it can also stand for simplicity. In his book *Square One: The Foundations of Knowledge*, the philosopher Steve Patterson, writes: "The Rubik's cube is analogous to the universe. Everything in existence starts in a solved state—every thing is how it is—and what appear to be paradoxes are merely scrambles of the Cube. With careful reasoning, all the scrambles can be solved—i.e., all paradoxes can be resolved. [. . .] The philosopher's job is to figure out how the Cube works—to grasp the basic concepts involved in a paradox—and to sit down and unscramble things until they make sense again."

Once, I thought of putting together a Cube book, a massive, visual compendium of all its many appearances. I didn't know what I was thinking! It became a never-ending task; there was always something new and startling. I started it not because I wanted to celebrate myself; it was simple childish curiosity from my part, and as a byproduct, I thought it might serve as nice stuff in an exhibition about the Cube's impact. No matter where I went, and I travel a great deal, the Cube pops up to greet me, turning an exotic and even slightly disorienting location into one that feels reassuringly familiar. Once I went to Phuket for a holiday, and in a small shop on a narrow side street, a place that seemed off the usual path for tourists, there was the Cube.

Another analogy: Seeing the Cube in all of these unexpected places is comparable to being in a foreign city, maybe a place like Kyoto, and hearing someone speaking Hungarian. I suddenly feel alert with recognition and the potential for connection. The same thing happens with the Cube, although on a far greater scale. It becomes a point of recognition, orientation, while at the same time, once you begin to become aware of it, there is a kind of ongoing vigilance. The Cube is not ubiquitous in the strictest sense of the word, of course, but no matter where you go in the world, it asserts itself. Here it is under the Brooklyn Bridge, and there it is in a Slovenian ski resort where a kid is working on it. Small moments. Quite incidental, really. But constant. Maybe if Ringo Starr hears

"Yellow Submarine" playing in an out-of-the way market, he has a similar experience.

ONCE, IN A SMALL COFFEE HOUSE IN SPAIN, where there were Cubes all over the place, on shelves, on the counters, on the tables, my wife and I sat down, and I picked one up and solved it. The waiter became enormously excited that I had accomplished what most other guests had failed to do. I didn't identify myself, of course, it made no sense given the setting, but I thought of how many times over the years I had glimpsed the gentle ways the Cube insinuated itself into the lives of so many different people. The waiter in Spain and the store owner in Phuket would have nothing to say to each other under normal circumstances; language and cultural barriers would seem insurmountable.

And then appears the Cube, bridging so many differences by a common point of connection.

As I consider its impact, I cannot ignore the toy market. When the Cube craze began, puzzles were hardly considered worthy of being toys at all by the mainstream. But the Cube's success spawned a whole new marketplace of twisty puzzles, a huge extended family with many relatives having the DNA of the Cube's principles of construction. There were two broad groups. The first, which includes the Cube, is based on the ninety-degree coordinate system. The angles of the three axes of rotation are right angles and the structure does not depend on the outer form but on the surfaces, which rotate on each other. If we preserve this regularity and then alter the number of the moving elements, we can get theoretically an infinite number of possible versions. The smallest one is a 2x2 with four elements, but then there are others with many more—a 33x33x33 made in 2017 holds the

Guinness World Record for "the largest order of Rubik's cube." It consists of 6,153 movable parts. But in the virtual world, there are no limits. I have seen a 100x100x100 cube on YouTube!

Another group of possibilities cuts up other regular solids in accordance with their symmetries, and applies our construction principle to them. For the tetrahedron and the octahedron, we can use four axes of rotation; for the dodecahedron and icosahedron, six—the number of parallel cutting planes then determines the number of moving elements.

Expanding an idea doesn't mean creating an entirely different object in a mechanical sense, even though there is often a challenge to make it workable. Each new iteration of the Cube was designed to achieve different goals of the users. Speed eventually became one of the most important elements. As speedcubing became more and more popular, the cubers needed equipment with which they could move even faster. The symmetries of Platonic solids, their richness in proportions and balanced potential are what makes it all workable. And yet, no matter how many different attempts there have been to alter it, no one could improve upon the Cube's original form, shape, and size. All of its characteristics came together in a distinct identity with a very strong character. And that is why, I think, it can mean so much to so many. This tension between simplicity and complexity, between the tactile accessibility of the object and what sometimes feels like the inaccessibility of the solution, and the fact that, in the end, the best way to approach the Cube is not like a task to be mastered, though that happens, but as an object to be played with—these are the provocative contradictions contained in what began as a wooden model so very long ago.

OF ALL THE QUESTIONS that the Cube inspired in me, none of them had anything to do with sports, competitions, or Guinness records. It would have been inconceivable to me at that time that such a global phenomenon would occur.

Not for the first time in my life was I completely wrong.

For it to become a real sport, "speedcubing" needed standardized rules and regulations so individual records in solving the Cube could be authorized and registered. A global community of twisty-puzzle enthusiasts already had started to unfold in the wake of the craze. However, it was the emergence of the internet almost twenty years later that brought the dispersed cubers together in a more structured way. Back in 1999, we built *Rubik's Games,* a suite of cube-themed games for PCs. It probably came too early, and it wasn't a great commercial success. Regardless, it provided a small place on the web where people interested in the Cube could start finding each other. Soon thereafter, a few hard-core speedcubers (most notably Chris Hardwick in the United States and Ron van Bruchem in the Netherlands) established email groups where people could submit their so-called PBs (personal-best times).

These platforms became the launching pad of the World Cube Association, a self-governing global organization with the mission of having "more competitions in more countries with more people and more fun, under fair conditions." Its inaugural event was the 2003 Rubik's Cube World Championship in Toronto, which closed the gap of twenty-one years since we hosted the first such event in Budapest. Since then, the WCA has organized a world championship event every two years. More important, with delegates in more than a hundred different countries, the WCA oversees more than a thousand local and international competitions annually.

In hindsight, competitions were inevitable. But somehow,

they came as a surprise. Typically, the key elements in competitions are either speed or some kind of quantifiable excellence of execution—you can see the distinction between the way that swimming and gymnastics are scored in the Olympics. For me, problem-solving is much more interesting than doing something that can be precisely measured. But I am in a minority. If you can do something quantifiable, people will eagerly try to beat the record. Or watch others try.

When a game is like chess and two opponents face each other, then the competition is obvious. But the essence of playing with the Cube is totally different. We are alone with the problem. Our opponent is the Cube itself and all its complicated nature. We are not faced with another human mind, so we cannot beat anyone at this particular game. What we are pitted against is *the nature of the problem.* If we lose, only our self-confidence suffers, but the next time we make a run at it, we may succeed. And yet competitions can be organized about nearly anything: How many hamburgers one can eat is just one crazy example.

(No crazier, I suppose, though, than some of the wild varieties of competitions that have emerged involving the Cube. As I have mentioned, there are astonishing records for the fastest times in which a person has solved a Cube while blindfolded, while riding a unicycle, while blindfolded *and* riding a unicycle, while juggling or skydiving or wing walking, and records for the most solved while running a marathon, or in a single breath underwater.)

But of course, the most sought-after record is simply for the fastest time in which a person can solve a 3x3x3.

Speedcubing competitions began in the early eighties, including national championships organized by the distributor, Ideal Toy.

As with running or skiing or skating or driving, mere fractions of seconds divide winners and losers, and small details become very significant as well. So it is with speedcubing. And as with these other competitions, equipment plays an important role. Speedcubes have a different feel than conventional cubes. A very important factor for speedcubing is the speed and precision of turning around the axes. If your turn is not complete and you didn't get back to precisely the alignment of the planes, it can blow up the whole thing. Or maybe one just can't go any further. When you pump up your bike wheels with high pressure, they become very hard. If there is less pressure, they become soft. There is an optimal pressure given a biker's style or environment. Similarly, professional cubers appreciate the effects of softening or tightening the movement on the axes.

The audiences at speedcubing events are especially intriguing for me, because I am not sure if they even know what they are looking at. A stopwatch, a blur of hands, and reds, greens, yellows, whites, oranges, blues, blacks, and then . . . it's over! It is an odd spectator sport, and yet, there is drama.

I WAS RECENTLY IN ATHENS for a science festival and the Greek national speedcubing competition. When the competition was over, at the gala that evening, I gave a speech (never my favorite indoor activity) in which I tried to explore a bit the significance of speedcubing. It seemed oddly appropriate to do this in the home of the original Olympics.

One obvious aspect of these races, I said, is connected to what you think about time, since speedcubing is based on the premise that doing something faster is better. This approach is well suited in the computer world, where speed is a god. The task is very straightforward and whoever is able to finish the task in the shortest time is the winner. And yet, in reality, time in itself is uninteresting. When we are faced with problems or tasks, we have a requirement, or just a personal need to do the job, make the effort, and hit the target without paying attention to how much time it takes. Simply arriving at the peak of Mount Everest is in itself an admirable achievement.

Timing becomes important as a clear indicator of amazing achievements you can't see. A cuber can only be quick if they have a deep understanding and knowledge of hundreds of algorithms and excel in pattern recognition, procedural memory, not to mention being a maverick with their fingers.

Every competition involves winners and losers, but there is no need to attach any great importance to losing. Most of the winners were losers before and vice versa. A few ex–world champions of cubing often appear at these events, not to reclaim their titles, but for something different. After all, to have been a world champion is by definition a temporary existence. As I said before, in a different context, the primary route to success is failure; you learn much less from success. But in the world of speedcubing, success and failure do not define the experience.

What drives people to participate, what becomes a real centerpiece and magnet for these competitions, is the community itself. The whole atmosphere revolves around people who love playing with the Cube. Often, they have already progressed through the smaller local competitions, which are open for anyone to participate. Small events usually involve no one but the cubers. The bigger the event, the more media appear.

But this is not the Tour de France, where you can follow the competitors as they make their way through the wild and adventurous landscape, the winding roads, the cheering crowds on the roadside. No one can really follow the cubers as they compete, however many people watch them. In some of the larger championships, the image of what is happening appears on a large screen above the stage, and when it is all over, the moments are played back in slow motion, because everything is happening with such a great speed, and it is too complicated to follow. But when it is slowed down, experts can analyze where someone made a mistake.

There are some sports in which the real opponent is nature and its forces, mostly gravity, against which the athlete pitches his expertly applied physical strength. In this sense we may say that the weight lifter's real opponent is not other competitors but the weight itself.

Over the years, the World Cubing Association developed very specific rules for how their events are organized: five rounds with five different starting positions. Each contestant's best time and worst time are taken out, and their three remaining times are averaged to produce their score. Fifteen seconds are allowed for studying the position without making a single twist before the performance, then the Cube has to be put back on the table on a sensorial plate. The players pick it up at the start signal, which triggers an electric timer. When they have solved the Cube, they replace it on the plate, and by doing that they stop the timer.

The current times are unbelievable, judging only by the standards of a couple of years ago. At the American International Toy Fair in 1980, my average time was more than a minute. By the time of the first international competition in 1982, the winner did it in 22.95 seconds in the second round.

His name was Minh Thai, a sixteen-year-old American who became the first World Cube Champion from among nineteen competitors who came to Budapest from nineteen countries as national champions. Now in the speedcubing universe that time is laughably slow. Here is what happened in a fifteen-year period for single-solve records: In 2003, the time went down to sixteen seconds; 2007 was the first time the Cube was solved in under ten seconds; then three years later, under seven seconds. In 2011, the shortest time was 5.66 seconds. And in 2015, the record dropped to below five seconds for the first time. The world record from 2018 was 3.47 seconds by Yusheng Du, during the Wuhu Open in China.

6

Life would be infinitely happier
if we could only be born at the age of eighty
and gradually approach eighteen.

—MARK TWAIN

WHAT DOES IT MEAN TO BE CURIOUS? I think it is the *capacity to be surprised and then to try to comprehend what it was that may have surprised you.* One gets curious about something that one finds unusual. It is an almost paradoxical fact that the deepest and most difficult questions can be found in connection to the most common things that we at first take for granted. To find the secret of a magician's trick is much easier than to understand why the apple is falling from the tree.

Curiosity is like thirst or hunger, a drive to fill a gap, scratch an intellectual and emotional itch. With curiosity, you have the feeling that something is there but you aren't quite sure what it is. There is an internal drive to follow up and some kind of promise that there may be a discovery of something

 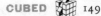

that is hidden and I may be the only one to whom it is not visible. What is beneath the surface is the need to comprehend. In that sense, we are not creators but discoverers. Inside that stone there is a statue; you need a sculptor to show you how it emerges.

Nature by itself, for example, is much more interesting than anything that can be imagined. We can't say we understand why the feathers of one bird are yellow and the others gray, how lightning bugs appear in some places but not in others. The best that we are capable of is spotting something as curious and then naming it. If there is sudden darkness in the sky, we are motivated to formulate a question. Why did this happen? Is a storm about to hit? Will it affect only my neighborhood or others? If you formulate them in the right way, many questions contain some answers as well.

Creativity can always be expressed in a question: What happens if . . . ? I like the expression "thought experiment" because it is such a precise way to characterize our imagination. How was it possible for the Greeks to measure the radius of the Earth in the second century BCE? They used the power of their minds, and Eratosthenes of Cyrene, who was a librarian in the great Alexandria library, had the inspiration.

He was the first person to calculate the circumference of the Earth and the tilt of the Earth's axis, and he never left his post in Egypt. He did it by knowing the position of the sun both in Syene and in Alexandria. He knew that the two cities were about five thousand stades, or roughly 673 miles, apart and noticed that when someone looked down a well at exactly noon in Syene, his shadow blocked the reflection of the sun on the water. Using a vertical rod, he measured the shadow in Alexandria at the same time. And by figuring out the angles and multiplying by fifty, he came up with the idea for the circumference of the Earth that was amazingly close to what we know today.

The ability I admire most in humans is the capacity of our minds to create abstraction from reality and use that abstraction to inform reality. It's what Einstein did with light. What Eratosthenes of Cyrene did with geography. In a way, it's what many artists have done with the Cube.

There is an old biblical phrase: "There is nothing new under the sun." Every new thing has elements of the past, but what is created is a new combination. I didn't create the Cube as a form. I didn't create the cubes that were put together to form the final product. I didn't create turning something around an axis. The wheel was one of the greatest and first innovations by humans, and soon it was discovered how the wheel made pulling weight easier by reducing the power of friction. Beginning with a rolling log, they discovered that it was possible to slice the log into cylinders, put two of them on an axle, and then we arrived at the electric cars we are using today. All that I did was create new relationships among these familiar elements.

Why? Simply because we remain curious all the time.

THE CUBE EMBODIES the mathematical principles associated with symmetry, transformation, and combinatorics. So it should perhaps come as no surprise that mathematicians were among the first people outside of Hungary to embrace it. David Singmaster, then a professor of mathematics at the Polytechnic of the South Bank in London, first chanced upon the puzzle at a conference in Helsinki in 1978. Several mathematicians had already gotten their hands on it, and Singmaster saw enough to become intrigued. He managed to obtain his own Cube from a Hungarian educator who had brought an entire bag of them to the event.

It was a serendipitous encounter.

After spending an entire night playing with the Cube, Singmaster would go on to become one of its earliest and most persuasive promoters.

Before it was widely available abroad, Singmaster personally sold thousands to colleagues and others; he developed what would become the standard notation for describing the movements; and he helped draw the general public's attention to it with lectures and articles.

Singmaster and fellow mathematician Alexander Frey have observed that the Cube "gives a unique physical embodiment of many abstract concepts which otherwise must be presented with only trivial or theoretical examples." In other words, when a layperson plays with it, they experience the laws that govern it for the first time; but when a mathematician handles it, they see long-familiar concepts finally brought to life. Perhaps the best example of the capacity for making concepts concrete lies in the Cube's relationship to group theory, a mathematical subject with applications ranging from art to physics, cryptography to card tricks.

It quickly became clear that solutions to the Cube could be systematized through algorithms—and what a rich universe of algorithms there were!

I had some personal experience in working with this after I got my first computer in the eighties and started to figure out how to program it. Not in a professional sense, of course, but I was interested in the thing itself.

To be in the digital world as opposed to the old, analog, "real" world requires a completely different way of thinking.

ERNŐ RUBIK

Everything can be created from one and zero, the switch is either on or off, and after that, all else follows. The magic is that it is so simple, and because it is so simple, the result is extremely complicated. I could never have anticipated how perfectly the Cube would become integrated into the digital era. The Cube immediately lent itself to pixel art (mosaics); it inspired robotics and challenged artificial intelligence.

One of the hard lessons anyone who has worked with programming learns is that if you are writing a program and make the tiniest mistake, one letter, one plus or minus, the whole thing just doesn't work. The consequences are profound. And the same dynamic holds with the Cube. If you miss one element of the routine—I prefer to call it a routine rather than an algorithm—the whole thing goes nowhere. Indeed, you lose everything. Not only did you not accomplish what you wanted to, you made things worse by losing everything that you had done before. And then you must start from scratch.

For me, the apps for the Cube represent a particular encounter of virtual and real "reality." I found more than 2,600 apps on many platforms such as Linux, Mac, Windows, and Android, most of them for free. It is very exciting to watch how IT people were inspired by the challenge. I can imagine the millions of users, holding their phone in one hand and the Cube in the other.

THERE ARE 43,252,003,274,489,856,000 (that is 43 quintillion, 252 quadrillion, 3 trillion, 274 billion, 489 million, 856 thousand) possible positions for the Cube to be in. Only one of these is the *starting state*, with all six sides in their own uniform colors. If we choose any of these positions, we can say that there is a certain distance between that one and the original state, measured by the necessary number of movements or steps or twists or turns that take us from the one to the other. That step can be used as a unit of distance, just as the yard was probably developed as a linear unit from the length of a human step.

Imagine lying on the grass on a beautiful, clear summer night, staring at the sky. There is very little ambient light, so the sky is filled with the stars that are never visible from a city. Randomly choose two. Then try to imagine attempting to travel from star to star. If your two chosen stars are next to each other, you need only one step. If there is one star in between, you need two. And now you have to get from one at the far end of the galaxy to another. How would you figure out how to do this with the fewest number of movements? This is close to the challenge that is contained in this question.

The sheer enormity of the problem hasn't stopped mathematical and computing professionals from trying to figure out the answer—indeed, a problem of this dimension is exactly what those who work with supercomputers enjoy. The method that they use is called "brute force"; some just call it "proof by exhaustion," and this refers to a mathematical proof that involves a highly sophisticated method in which each possibility is split into a finite number of other possibilities. Those possibilities are, in turn, checked and split to see if the proposition can actually hold.

The capacity of Google's computers and the brainpower of a group of mathematicians finally determined the smallest number of possible moves necessary to solve the Cube out of all the possibilities. The number they found is known to most people as twenty, but in Cube jargon, these two modest digits are now "God's number" of the 3x3x3 Cube.

Way back in July 1981, Morwen Thistlethwaite proved that, at most, fifty-two moves would always be enough to get from any one point to any other state. In other words, there would never be any two positions that would require more moves than fifty-two in any direction.

About thirty years later, in July 2010, computer programmer Tomas Rokicki started looking at the challenge with his colleagues. He built on the work of another mathematician, who had divided the task of solving the Cube in two different steps. The steps were divided into two configurations that were partially solved, and the number of potential permutations was not in the quintillions but was 19.5 billion. Researchers found that, using this strategy, a Cube could be solved in at most thirty moves. Rokicki grouped configurations together by using the special partially solved configuration set, and that meant solving an incredible 19.5 billion configurations at once (which is still so much less than 43 quintillion). On their website, cube20.org, Rokicki said they "partitioned the positions into 2,217,093,120 sets of 19,508,428,800 positions each." By working with 2.2 billion problems, rather than the original 43 quintillion, and using a bank of Google supercomputers, Rokicki and his colleagues proved that God's number was in fact twenty. That is, any particular position of the Cube can be reached from any other in no more than twenty moves. In this sense, one move means to make a turn around one axis—it can be either a quarter turn clockwise or counterclockwise or a half turn. If you are counting the quarter turns, this number is twenty-six.

All the extremely complicated and theoretical work by the mathematicians illustrated two evolutions in the life of the Cube. The first is their interest marked a new way in which the Cube was taken very seriously. And this leads to the second point, the extent to which this work shows the hidden structure of the Cube. Symmetries and group theory are intimately connected to the Cube, and mathematicians took the content seriously and worked hard on it. Their discoveries led to the Cube being used for cryptography—devising passwords, for instance. (In a way, for me, this does not have very positive connotations about people, suggesting as it does that we very much would like to, or need to, hide things.)

There is an official event at speedcubing competitions called the Fewest Move Count, in which competitors are judged with no reference to time, or even without interacting with a Cube! This is a pen-and-paper exercise in which cubers have to figure out the shortest solution from a given scrambled position. The number of winning moves are usually very close to twenty. (In speedcubing proper, cubers are generally far off from the theoretical ideal, and winners take around fifty moves.)

In the end, of course, a pattern is a pattern and, in a mathematical sense, there isn't a more or a less scrambled state.

TIME NEVER COMPROMISES. Our perception of the speed of time isn't at all stable. My grandchildren mark the days on the calendar before their birthdays with large Xs; the time just drags until their celebrations. It makes sense, of course, because as a proportion of a five-year-old's age, two weeks is far greater than it is for a man in his seventies. Naturally, it would feel as if it passed much more slowly. If a short period of time is crowded with many things, it can seem as if it all took place over a very long time. I look back on the wild and intense early days of initial Cube craze and can't quite reconcile everything that happened with the relative brevity of the time during which it all occurred.

How we see time shows the differences between knowledge and feelings. Objectively, it is all the same, but we feel the differences.

I consider the blur of time that has passed with my now middle-aged sidekick; we have been companions for nearly fifty years. And what a half century it has been, with so many momentous political and technological events and transformations. The speed of technological and scientific change between 1974 and today cannot be compared to any time in human history.

As we lived through it, each change seemed so gradual as to appear almost unnoticeable, but then we look back on it and see the significance of its impact.

When Steve Jobs introduced the iPhone on June 29, 2007, few people outside of the tech industry—and possibly not even them—could have understood the utter transformation of our lives that it suggested about the future. Mobile phones were already owned by nearly everyone, from schoolchildren to grandparents. But who on earth would need to carry around a mobile computer, not to mention one that also carried more music than anyone could listen to in a lifetime? As we all know, everyone soon felt they needed such a product. (Unsurprisingly, I was a late adopter. I had a simple cell phone for years, and when people in our company would replace theirs, I would take one and give it a new battery and change the SIM card and use the phone until it wore out. After years of doing this, all the simple flip phones disappeared, and I just had to get a smartphone.)

In a mere five years, a heartbeat in terms of civilization, cutting-edge technology became part of daily life, and we have been profoundly transformed by it. And yet, we remain the same. I am reminded of the science-fiction books that I loved as a teenager—and still love—in which the authors seemed to have anticipated the speed of communication, the astonishing genetic work, and the highly personalized technology that we enjoy today.

In those books, the scene was usually outer space, not here on Earth, and they often looked forward to the far distant time of the twenty-first century, imagining that after the year 2000, we would occupy some kind of dreamworld.

However, the genre of science fiction is very different from that of fantasy, even though they are often included in the same breath. Sometimes books combine the two, but they are very different. The qualities that define fantasy—mythical beings, characters with special powers, and unrecognizable settings with talking animals and plants—have remained quite stable throughout the ages. What was fantasy in the past will remain so in the present and the future—fantasy is created by fractions of reality, but without any real connections to the laws of nature. Consider a dragon: He has wings and can fly without feathers. He can breathe fire. The elements of the fantasy creatures are all real, but the way they are put together is impossible. When we mention science fiction, what does the science mean there, and what does the fiction mean? The combination of the two terms speaks about things that don't yet exist but are not in contradiction with the laws of nature. One hundred years ago we could speak about spaceships and rockets. They may not have existed, but in keeping with the laws of physics, they were always possible.

Some of the changes science fiction authors imagined were more radical than those that have really happened. They described travel through time and space that we have not yet approached. Yet they were unable to dream of many things that today are considered to be trivial or are taken for granted by my grandchildren. Even the science-fiction writers whom I read did not fathom that at some point, instead of a bedtime story, children would simply be able to search for their favorite videos on YouTube. Often in those old science fiction narratives, the problems on Earth—the poverty, inequality, conflicts—had been resolved, and the remaining challenges and possibilities existed on other planets or in other galaxies. There was a kind of optimism and confidence in science and technology as a way to resolve social and environmental problems and political tensions. There was no recognition of how serious, nor how dangerous, environmental problems would become.

And what they could never have fully appreciated was how so much more has happened in the last five decades than has happened in centuries of human history. I am not just speaking about great leaps in scientific discoveries or political changes, but in what is more interesting for me to observe: the dramatic changes in the daily lives of human beings. So many new jobs and professions have emerged. And many of the old reliable ones have disappeared. For my grandchildren, the world is much more open than it was for us. They have already seen more of it than I did before I was thirty. Their English is strong, and their futures will be very different from any that I can imagine, much less experience (though I can't say that it will be easier).

What kind of different societies will emerge? Will they be utopias or, what is very fashionable now, dystopias? In my childhood readings, utopias and dystopias were both portrayed. Today, if you are reading about the future, most futures are dystopic. The ultimate catastrophe is caused by many things: germs, AI, the invasion of aliens, the collapse of the environment. But all these threats are used to express our fear. This is not fantasy. When in a laboratory people are working on bacteria or viruses that can be weaponized, the stories are not in contradiction to physical laws nor the laws of human existence.

In my lifetime, we have moved from a period when there were limited choices for most people, to a time when there are so many choices, it is difficult to know what to pay attention to. That would not be easy even if all the information out there were true, but it is more difficult when so much of it isn't. When I think about my grandchildren, their main challenge will be to find the right information in order to decide what they should do. But to make them capable of doing this will not be easy. What do they need to be equipped with in order to be capable of determining what is false and what is true?

In the past, people considered themselves to be the center of the universe, and everything revolved around them. They would anthropomorphize the natural world and the inanimate objects around them so that they imagined they had power over, or perhaps a personal relationship with, the rising and setting of the sun.

The twenty-first-century counterpart of this worldview is evident in our approach to artificial intelligence. Siri and Google and smartphones all utilize artificial intelligence. We have gotten so used to interacting with them that we barely appreciate how remarkable it is to be able to tell an object to select some music for us or offer the shortest route to a restaurant, or that it replies appropriately when we ask it a question. We should be awestruck by this, and yet we aren't anymore. The warp-speed evolutionary drive of technology has all somehow been seamlessly integrated into our daily lives.

Yet, throughout the experience of moving through time and witnessing radical changes, we nonetheless have maintained a stable identity, an unchangeable core. How is it possible to remain the same, when everything around us is so dramatically altered? Should someone be satisfied and accepting, or critical and often uncomfortable? For me, I believe one must be both all the time. If one is constantly satisfied, then they are unlikely to be willing to change or even adapt. If one is too critical, it becomes impossible to enjoy what you have. Where does one find a kind of fixed point?

AI is interesting and important and dangerous. The danger is not in the technology, but inside us. We should first look in the mirror.

Our history shows that if we would like to do something, we always seem to find a way to do it more quickly and more efficiently. This is part of human nature, the enjoyable experience of something working efficiently without any serious consequences. Most creations by humans are value-neutral: One cannot say if the object is good or bad. But there are two omnipresent dangers: the object's potential and the user's intent. We are the ones who make an object good or bad.

In an episode of *SpongeBob SquarePants*, the Cube is presented as transcending the skills of humans and becomes synonymous with "artificial intelligence." It becomes a kind of litmus test for measuring exactly how intelligent, imaginative, and sophisticated AI is by demonstrating how quickly and cleverly it can learn to solve the Cube independently without human help. While so much of AI work is complex and theoretical, when it is connected to this three-dimensional puzzle, there is a measurable and self-contained point of reference.

It is the same thing with the terribly complicated and efficient robots that can solve the Cube faster than any human. These robots have created and then surpassed one record after another. One scholarly paper by four professors from the Department of Computer Science at the University of California, Irvine, was entitled, "Solving the Rubik's Cube Without Human Knowledge." They taught a computer to solve it with something they called "reinforcement learning," in which the algorithm learned a policy that determined which move needed to be made from any given state. In just forty-four hours, they trained a network to solve randomly scrambled cubes, which they dubbed Deep Cube, and then compared its performance with two other solvers.

One hundred Cubes were given to this computer, and they were all solved under an hour. Scientists evaluated the number of combinations that the Deep Cube used in comparison with another solution. What they found was that machine-learning algorithms generally used pattern recognition, not reasoning, to solve the problem. By combining neural networks with AI that had been programmed to work with symbols, AI became capable of using knowledge to solve a particular problem, but it did so only through being reinforced in its moves. That is different from robots playing chess, when one move is less promising and another has more potential for the future.

The Cube is not so obvious: You can make several promising moves, and even those that may appear to be wrong can lead to some interesting solutions.

When the paper by these researchers was published, there were many articles about their achievement. So far computers had been very proficient in games like chess and go, but solving the Cube was much more complex. The way that AI learned how to play and win these board games was through a system of positive reinforcement. When the machine made the correct move it was rewarded, so it learned what was involved in playing the game successfully.

It's much more complicated with the Cube because there are so many possible moves, one can't really tell if a single move brings one closer to the ultimate goal.

What these researchers managed to do was to break through the challenges by letting the machine learn how to evaluate the move all by itself. Amazingly, it starts each move by comparing its present state with a finished Cube and working backward to assess whether a single move would bring it closer to the target or not. In June 2018, the *MIT Technology Review* announced their accomplishment, saying, "Yet another bastion of human skill and intelligence has fallen to the onslaught of the machines. A new kind of deep-learning machine has taught itself to solve a Rubik's Cube without any human assistance."

Robots have two basic parts: a "brain" and a mechanical existence. When researchers tried to create a robot that was capable of walking like a human being, they discovered the complex details involved in the process of walking. It is not by accident that we aren't able to walk immediately. We need time to learn how to coordinate many parts of our bodies and learn to balance while making very complicated movements, and that does not even factor in how we respond to the environment, whether we are climbing the stairs or moving quickly to avoid a pothole.

When robots were introduced to the Cube, one of the most complicated initial tasks was also one of the most essential: teaching them how to make a turn.

I have seen a video about an artificial hand playing and have admired how cleverly the hand moved as it turned the Cube. The other challenge involves designing a program that is capable of directing the physical machine with the goal of solving it. After all, in order to do this, the machine needs some capacity for perception. Every Cube at any point has a pattern, and the pattern determines at what stage the Cube is toward a solution.

Robots have been built with the special expertise of solving the Cube. A RuBot was made with a robot head and robotic

arm; it spoke and walked and solved the Cube very slowly. This was within the realm of specialized, goal-oriented robots that were capable of building cars, as opposed to universal robots, which are made for several tasks. During a running for records between specialized robots, the actual world record holder managed to do it in under 0.4 seconds! It's just a blink, you can't see what happened, and it is finished. You need it to be slowed down by factors of thirty to even see on the screen what is going on.

IN ALL OF THIS, I AM REMINDED of Isaac Asimov's 1950 book, *I, Robot*, which was a collection of short stories about robots. In one of them, he described "Three Laws for Robotics" that would be very useful to apply to AI, especially the superintelligent version, which is being developed in labs and universities and government offices and industry all over the world.

The laws are: 1) "A robot may not injure a human being or, through inaction, allow a human being to come to harm"; 2) "A robot must obey the orders given it by human beings except where such orders would conflict with the First Law"; 3) "A robot must protect its own existence as long as such protection does not conflict with the First or Second Laws."

Replace "robot" with "AI" and one begins to see some of the challenges and the need for these constraints.

The expectation, it seems, is that the artificial intelligence we are creating in the far distant future will resemble us. How could it not? The computer that first solved the Cube did so by employing a number of human qualities, like learning and independence and rational "thinking." But doesn't that mean that the fear we often have about AI and its implications also reflects a fear that we have about ourselves?

If artificial intelligence becomes the same as ours, we are in trouble. But, all the time as parents, we hope that the next generation will be more clever than we were, and they will be able to solve the problems that we created. Maybe they will even be more successful than we are, who are still suffering the consequences of the mistakes made by our predecessors.

WHEN I IMAGINE THE CUBE, I see a structure in motion. I see the framework of its edges, its corners, and its flexible joints, and the continuous transformations in front of me (before you start to worry, I assure you that I can freeze it anytime I like). I don't see a static object but a system of dynamic relations. In fact, this is only half of that system. The other half is the person who handles it. Just like everything else in our world, a system is defined by its place within a network of relations—to humans, first of all. In design, the method breaks systems down into modular elements and considers how they can be disassembled and reassembled in completely different ways.

It's a fundamental approach to the modern world. The word "system" has many meanings, but the one that I find most useful was expressed in the reliable *Collins English Dictionary*: a system, it states, is "a group or combination of interrelated, interdependent, or interacting elements forming a collective entity." By taking seriously each part of this sentence, how an entity is "interrelated, interdependent, or interacting," it is easy to see why this would appeal to me, and why I think it would relate to the Cube.

A system encourages us to look at the big picture: How does an individual element fit? If something has structure, then something else keeps it together. We consider structure to be physical, but it is also about how all the elements come together and interact.

A systems approach means paying attention to one element—a creature, a plant, a river, a family, the Earth, a white cubie—*as part* of something bigger that is formed by smaller entities. But it also means looking at the various components that create a single entity. You can divide a system into elements, and it is usually described as an interaction of elements both outside and inside.

You are probably already familiar with one popular ambassador of systems design: modular furniture. Even if modular furniture is something you have never assembled or experienced firsthand, it is a good way of explaining how this works. Shelving that can be combined in several unique combinations and sofas that break apart into individual pieces are both examples. It may not surprise you to hear that I also once designed modular furniture systems. The fact is, however, that each part of modular furniture is both autonomous and connected to other pieces.

Systems appear all across the map of human activity, from economics to information technology, from business management to, of course, design. But despite all these uses, I believe that systems thinking has even broader implications than those that we usually recognize. At the most basic level, creative change is driven by the rearrangement of elements from existing networks. Recognizing the richness of this perspective and mining it for new solutions is an invaluable asset in the creative process.

THE SYSTEM OF THE CUBE is really three systems: the constructional, the functioning, and, the one most people experience, the interactive. One person interacting with one Cube.

The Cube's construction is a closed system with its assembled elements all contained. But once someone starts to play with it, it becomes an open operational system, because it can't move by itself. The system is rather small—there are not that many elements—and not hierarchical: The elements are distinct and individual, but no single one is more important than the other.

Each element of the Cube shows only its individual capacity to function and nothing more. With as many sides as it has colors, as many colors as orientations, its orientation conforms to its form and function. Corners have three sides, three colors, and three possible orientations; edges have two sides, two colors, and two possible orientations; and middles have one side and one color, and thus show only one orientation (although in reality the middles have four, which would have been rendered visible by a different kind of coding).

It is a more complex system than you might have imagined, isn't it? No single characteristic is more important than the rest. It is when they all come together that something remarkable occurs.

But that requires someone to play with it. Each individual is an extremely sophisticated system, possessing both physical and mental capabilities and their respective limits. There are the motor movements of hands and fingers and visual perception. In this system, the most important role is played by the brain, which interprets all that is taking place. But the brain creates its own difficulties by making assumptions, being unable to remember, losing its way. Many people who enjoy the Cube reach out to others who do, and a community is created.

When so many people engage in communities, interacting and exchanging knowledge, influencing each other and admiring what the other is doing, what happens? Yes, yet another system!

The Cube is not an inflexible system. But the key is to prioritize what is most important and maintain the basic structure. We change the small details all the time, we wear different clothes every day, but there is something that characterizes us, and to keep that character is the best we can do. And to be able to maintain that constant is, in itself, quite something.

BEFORE I ADDRESS THE UNKNOWABLE mysteries of the Cube, let me first reflect on what I most certainly know about him and what I still believe to be his most important mission. Since my name has inadvertently become a brand, I can most identify with it as an educational brand. I want "Rubik" to flourish by fostering students' curiosity and nurturing a lasting passion for complexity, creativity, and innovation.

During the eventful year of the Cube's fortieth birthday, in 2014, I was invited to Cambridge Union, the prestigious university's historic debating society. It was a unique and memorable experience, as I was joined onstage by leading mathematicians, scientists, and psychologists of global renown. We were discussing the surprising intersection of autism and talent. While it might be unexpected, children and adults living with even quite severe autism are sometimes absolutely brilliant when it comes to the Cube. (So much so that one of the world's very best cubers today is among them.) There might be many reasons, but one is surely their outstanding focus and unwavering attention to a challenging task. A child with autism may be less adept socially, even withdrawn from their classmates, but possesses a capacity to focus that many should envy.

While it is a wonderous thing to see someone who is otherwise marginalized to flourish with the Cube, the real excitement comes when we consider its educational promise in more general terms. My academic hosts in Cambridge later helped me to uncover the science behind this vast potential.

The Cube builds on cognitive and emotional skills that are at the very core of learning and succeeding in the twenty-first century. Solving the Cube requires and improves visual working memory (which happens to be a better early indicator of academic achievement than IQ scores). As one gets more familiar with it, like speedcubers, the Cube is also an excellent illustration of "procedural memory" in action, which is psychological jargon for our body and mind knowing how to do it.

Beyond the many other cognitive skills already mentioned (like spatial thinking, pattern recognition, sustained attention) and obvious motor skills, solving the Cube also speaks to our emotions. In the first instance, it teaches us to tolerate frustration! It is an improbably difficult puzzle, and rewards go to those who can appreciate the whole journey, not just the final destination. At the same time, exactly because it is so challenging, the gratification when success occurs is truly meaningful. Arriving at the solution boosts self-confidence by a strong sense of competence. This can be transmitted to other demanding mental tasks. Plus, eventual success is immediately recognizable by anyone, so the Cube invites social recognition as well—just remember Will Smith's character in *The Pursuit of Happyness*.

By all these characteristics and more, the Cube is a natural educator! Indeed, he was born as a teaching tool for my own students—who are now mostly grandparents, just like me. We have seen how the Cube feels just as at home with highbrow mathematicians, experts in group theory, geometry, and symmetry, as he does with specialists in engineering, robotics, or computer science. At the same time, the Cube's natural habitat is in the hands of young children who play and learn, all the way up to the teens and young adults who dominate as champion speedcubers.

"Edutainment" is a recently introduced concept that suggests that learning should feel like play or entertainment for kids—and their teachers should feel confident that they are in fact delivering education. (An old math professor friend of mine once joked that the big practical problem with this otherwise most appealing concept is directional. When children are enjoying the process of a learning task, it is the teacher who considers it entertainment; and when the teacher believes she entertains, the kids still perceive it as good old education!)

The Cube is one of the very few examples where edutainment is a seamless experience. The teachers happily accept it as a vehicle to teach and the students enjoy the playful learning the Cube can provide.

A decade or so ago, in the United States, there was an effort to reform the teaching of science. This resulted in what became known first as STEM, to cover four scientific fields (science, technology, engineering, mathematics), or, my preference, STEAM, which would also include the arts.

In this context, teachers took the Cube and began using it as an educational tool. The You CAN Do the Rubik's Cube curriculum has been developed for teachers with the goal of teaching children from kindergarten through high school not only how to solve it and experience the benefits of this accomplishment but to use it to learn a number of different concepts.

The curriculum helps teachers engage students in an interactive and tangible way while imparting learning about algorithms, the geometry of solids, ratio and proportions, mathematical operations, algebraic thinking, and even physics. In another module, students learn about Fibonacci, the code he created, and how the Fibonacci sequence relates to real life and the perfect spiral. Students practice drawing perfect spirals and learn how patterns are a mathematical concept that surrounds us. They can learn engineering by determining a problem and systematically figuring out a solution. One exercise is for them to figure out how to use the Cube to play tic-tac-toe.

Another great adventure in STEAM education for me was our collaboration with Google and the Liberty Science Center, one of the largest such institutions in North America. The interactive exhibition *Beyond Rubik's Cube* is all built around immersive learning experiences, inviting visitors to playfully understand sometimes bewildering topics from algorithms and problem-solving to model-building and inventing. The exhibition first opened in 2014 and is still on the road, traveling to science centers across the United States and Canada.

The Cube has become an extremely accomplished teacher, from kindergarten through graduate education. At each level, he manages to do what the great teachers have done since Plato: meet students at their level and raise them higher.

MYSTERIES ARE NOT JUST what we ourselves cannot understand, because that suggests that someone else could; that with enough persistence or help or learning, something is possible to comprehend. Given the growth of scientific knowledge, many phenomena that once were mysteries are now within the realm of comprehension, but many, thankfully, are not. There is the mystery of consciousness, of life and death, of love and art. Max Planck, the great physicist, once said, "Science cannot solve the ultimate mystery of nature. And that is because, in the last analysis, we ourselves are part of nature and therefore part of the mystery that we are trying to solve."

Planck's observation means a great deal to me. It is much easier to understand things from an outside view, with some detachment, than from the inside. So in a way, perhaps, I am least equipped to fully explain the mystery of the Cube.

The Cube contains many mysteries, but among them these three are especially compelling: the mystery of how long it has been around, how deep its emotional touch can be on an individual, and how broad the effect has been across the world. How could this idiosyncratic little object acquire such immense popularity and continue to thrive? Indeed, why was there a craze at all? And how has it not been consigned to the past but instead remained an active presence a couple of decades into the new millennium?

Mysteries are apparent in so much of our lives that we take for granted. If we are curious and have the capability to spot it, every corner we turn is filled with mysteries. Once we start thinking about them more analytically, once we go into greater depths to understand how they work or where their essence might be, their apparent simplicity is shattered. For example, we take language for granted until we start learning a foreign one. Then the vast mysteries of our mother tongue begin to become apparent—from basic language acquisition to the realization of how it reflects the more subtle and profound thought patterns of native speakers, and the rhythms and the deeper meaning of poetry.

Then there are entities that, although experienced as simple, are in apparent contradiction with our knowledge of the world. For example, the mystery of sight. Scientists have untangled a great deal about the biology and neuroscience of sight but still cannot explain the real individual experience of what it means to see for humans and other animals. They can explain its mechanism but not what each individual experiences when they look out the window or at a painting or, especially, in the mirror.

It is as if we go through a cycle: The more knowledge we accumulate about a particular natural or man-made phenomenon, the more it changes from something mysterious into something merely "complicated," then into something simple, until we probe and question the elements of its simplicity, and it enters the realm of mystery once more. Usually we call something "complex" when we cannot understand it. Complexity by itself is not mystery.

Some things appear to be extremely complicated at first sight. We immediately see that they are immensely difficult and composed of innumerable and very subtle aspects that we realize are beyond our comprehension. This is positive; it implies that something is valuable, invites understanding, and suggests the possibility of discovering new and concealed values.

Simplicity is much more mysterious than complexity because it gives the impression that everything you need to know is right in front of you. But it isn't. And then we begin to realize that often what we consider to be simple is extremely complex. Then the questions can begin.

FOR SOME DEEP REASON, FORMS EVOKE EMOTIONS in all of us. Obviously, at least in theory, one form is neither better nor worse than another. But somehow, once we consider them not in relation to each other but in their relation to how we respond to them, we like or dislike them; one attracts us, another repels us, one seems to be reassuringly familiar, the other strange, one appears to be friendly, another unfriendly.

Of all the other possible forms, there was the one I obviously gravitated toward. But why a cube? What is it about this solid little shape that resonates with me? The cube is an ancient form. It is one of the Platonic solids and has a very basic and fundamental character. Part of it is the ninety-degree relationship between its edges. The square by itself is very familiar and easy to recognize. And each of the sides is two-dimensional. The right angle is probably one of the first angles ever discovered. Standing, for humans, is one of the first human achievements, distinguishing us from the apes, and so, too, does the cube stand firm.

Consider the mystery of gravity. You want to plant a rod in the sand. If you do it vertically, it stays put. But at some other angle, it will eventually fall. The right angle has some kind of elevated distinction from all the others. You can see this in the right-angle edges of the Cube, which are neither sharp nor blunt; they are just, well, a kind of perfection. If pressed, I would say that I like how the Cube is defined by its angularity—unlike the sphere, which lacks such definition, since all its points are identical—and its regularity, which surpasses that of all other Platonic solids. You can mark the sphere, but it does not mark itself, which evokes an odd feeling of insecurity. The fewest number of elements suggests the highest level of regularity (or simplicity).

It is difficult to explain emotions; their many intertwined aspects defy straightforward analysis. One can communicate emotions successfully only to those who share or have shared them to some extent—with others who are on the same wavelength. As I describe this, I hope that everyone has experienced something similar in regard to the Cube, even without ever having been consciously aware of it.

AS AN ARCHITECT, SPACE has an orthogonal coordinate system for me, which is a fancy way of saying that all the axes meet at right angles. This is due to the fact that we live on Earth in a gravitational field whose force of attraction tends toward its center. The Earth is big enough to make us experience the curved surface upon which we live as flat, which is relative to the vertical direction of the gravitational force. (It would be amusing to write a science-fiction story about a globe that is so small, its inhabitants actually experience its curve.)

Architecture—and somewhat more crudely, the act of building—are struggles against gravity. One cannot eliminate it; all one can do is adapt to it, an experience that must have been one of the first intellectual experiences of *Homo sapiens*. In fact, all of us living on this Earth—plants, animals, human beings—do this. We have managed to adapt to gravity so that we no longer notice it; it is a fact of life from our very first, tentative steps. Only billions of years after the first tiny signs of life appeared on Earth were the cosmonauts capable of breaking free from this implacable force. Architects have to be conscious of gravity every minute, because it is the crucial factor in all our work, from the smallest detail to the largest structural questions.

And yet, this still leaves open the question of why some objects produce positive feelings in us, while we react negatively to others. Since we live in a three-dimensional world, surrounded by all kinds of forms, we recognize the objects surrounding us by their shapes—the angles of a house, the cylinders of trees, the circles of tires on our rectangular cars.

Even without our conscious acknowledgment, these shapes or forms carry emotional weight and significance.

IF ONE MYSTERY OF THE CUBE IS ITS SHAPE, another is its 3x3x3 identity. Numbers have stirred people's imaginations from ancient times and inspired their own branch of mysticism. Some numbers are considered lucky, for example, while others are unlucky. For me, the number three seems to have a particular significance, relevant in some strange way to the relation between man and nature.

Three is one of the prime numbers. In some cultures, three is the symbol of perfection. It represents the unity of the body, soul, and mind; the earth, the sea, and the sky; power, knowledge, and existence; or God, nature, and humanity. Pythagoras, the great philosopher and mathematician in the sixth century BCE, who first explained the relationship among the three sides of a triangle, believed in the tripartite structure of the universe and that every problem in the universe could be reduced to a triangle and the number three.

When I mention these historical references, I am trying to suggest that the number three has a kind of power. It reflects some hidden laws or structure in the universe. People often feel as if something is complete if there are three, because the number three represents some idealistic unit. The most stable chair or table has three legs.

In nearly every religion, three is important: The Holy Trinity is the central belief in Christianity. The Tree of Life is divided into three symbolic parts: the roots that represent beliefs, the trunk that represents the mind and body, and the branches and leaves soaring toward the sky that symbolize wisdom.

We live in three dimensions—height, width, and depth—

and time is divided into the past, the present, and the future. Ancient Egyptians also divided time into threes, and for them, a month was not divided into four weeks of seven, but three groups of ten days. In religious presentations, the Egyptians required three pieces of fruit, and they were buried with three stones. Believers made their offerings to the gods three times a day. In ancient Greece, Kronos—or time—had three sons: Zeus, who reigned over the heavens; Poseidon, who was in the ocean; and Hades in the underworld. In the underworld, Cerberus had three heads and the Oracle at Delphi sat on a three-legged chair. The length of a leap year is 333 plus 33 days.

We all know that the third time is the charm. The Hungarians believe that the third time tells the truth, and the French suggest that all good things come in threes. There is a Hungarian saying: "*Három a magyar igazság és egy a ráadás.*" Which means, "Three is the Hungarian truth, and one more."

And we shouldn't forget the Latin phrase "*Omne Trium Perfectum*": Everything that is three is perfect.

In the Cube's life, it took three years to get the first manufacturing completed. (I was 33 back then.)

It took three years for him to break through the Iron Curtain.

And the height of the craze lasted, yes, three years.

MOTION IS ANOTHER ONE of the joys and secret sources of the mystery of the Cube. The process of breaking the solid by movement is a bit like squaring the circle—not in the sense of the ancient mathematical problem, but in the sense of attempting the impossible. The angularity of the multiple-square form may seem impossible to rotate, and yet it is capable of revolving. The freedom of the turns is yet another source of disorientation.

Spinning and rotation have a mysterious appeal of their own; the potter's wheel is a perfect example of the way other shapes are created when we turn. We all can relate to the thrill of spinning. Children are mesmerized and delighted by a spinning wheel or a humming top. The Whirling Dervishes, who descend from the thirteenth-century Sufi mystic Rumi, spin with speed and discipline that seem impossible for mere mortals, but in doing so they believe they glimpse the divine. We all remember becoming wonderfully dizzy as we spun on crazy amusement park rides, going around and around until we staggered and abandoned ourselves to the magic of vertigo; we are attracted to all the controlled danger in that space. We might be toying with losing terrifying control over something core to our functioning in the world: our correct orientation and equilibrium.

This reaction to spinning may even be more primitive: linked to the way that the Earth spins and creates our days and our nights, and the way our blue planet travels around the Sun so we get the seasons, summer and winter. As soon as mankind understood this phenomenon, it marked the beginning of civilization, the first hint of apprehending the workings of the universe and the motion of celestial bodies.

In contrast, turning as a source of recreation is also common—one thinks of the roulette wheel, but that is merely a game of chance, like a throw of the dice. To spin is usually a full-body experience, while turning is more self-contained, more restrained. Turning something by hand can be especially satisfying; it is an activity that is harmonious with the way that our joints function. The capacity to turn is one of the really unusual parts of the Cube as a structure; it is what makes it seem alive as we recognize how much change can be the result of just a few simple moves.

Another aspect that makes rotating mysterious is that it is difficult to follow or imagine its final result. In Hungarian, we have a term to describe someone who has a very unconventional way of thinking: a "Csavaroseszű," which literally translates into "screwminded." But in fact, it has a positive connotation, suggesting someone who is not just clever, but whose thinking is very original, departing from the predictable to come up with surprising connections. As anyone who has worked with the Cube can attest, it often feels almost impossible to discern the logical conclusion of rotating sequences.

EVEN THE CUBE'S VERY APPEARANCE is mysterious.

The form, elemental to the extreme, reveals at first sight almost nothing about itself, nothing of the challenges it contains. It is not by chance that it has so often been described as "innocent looking," highlighting one of its most important contradictions. Some objects at first sight are as baffling as assembly directions in Japanese (for those who do not read Japanese), but the Cube in its calm state is dramatically simple. When all the colors are in place, it suggests peace, a sense of order and security. The regularity of its shape, the recurrence of identical forms, the tranquility of the planes, the compactness of the closed form are in sharp contrast to all that it means once it is brought to life, when it is in motion and changes.

Its movement is quite free and invokes fundamental human concepts: order, harmony, disorder, and chaos. I have come to think that its connection to these qualities is crucial. Especially since these terms are not objective but informed by our relation to the world surrounding us. We say "order" when we think of the law, but the connotation is also one of predictability. Order conveys a guarantee of orientation, like the alphabetical order of words, the order of the merchandise in a store, soldiers marching in close order, trains leaving the station on time, numerical order. When something is in order, we know that everything is in its proper place in space and time.

Order is certainly a valuable trait, but I don't believe it's enough. What we need beyond that is harmony. Harmony means the right proportion, with "right" containing the subjective element. It may be reassuring, upsetting, pleasant, or unpleasant, but even when there may be some dissonance, it is balanced by something harmonious. "Harmony" is typically understood in a classical sense, implying a measure of something idyllic, a lack of contradictions, a collection of individually autonomous elements that come together to create something new. For me, it is really no more, no less, than the right proportion in space and in time; this proportion is always defined by a network of interacting relationships. Its result can potentially be beauty and truth. Order can be very boring, but harmony is something special. We say something is in "harmony" because we can *feel* it.

DISORDER IS THE DEGREE OF DIVERGENCE from an imagined or expected order. It is possible to measure disorder, but chaos is immeasurable disorder. We have no point of reference from which some order can be imposed, and for that reason it is hopeless. Chaos has too many components. If there are only a few elements lacking in order, we may always find some way, some law, some institution that can be a key to impose order. When there are too many elements to calculate, we associate them with "chaos." And yet, so often we hold the chaos of the Cube in our palms. How surprising that such a small object, consisting of so few components, nevertheless evokes this trait.

The Cube's chaos is always born in the grips of order. Over and over again, its order is established and lost and reestablished in the same form. The moving, colored cubies create chaos on the surface for us, against their dark background. There is nothing hidden about all the changes and the movement. Everything takes place right in front of our eyes; we look at it, even attempt to follow the moves, but in the end we cannot see what is really happening, like a magician's sleight of hand, but this time there is no illusion. The order-break-order-order-break-order-order sequences have a cumulative effect, like a drumbeat that provides the rhythm to the music of the colors.

THEN THERE IS ANOTHER MYSTERY that few even suspect: the function of the twenty-seventh piece, the invisible core.

It is stable and never moves, acting as the point where the three axes cross. There will always be two directions from which a coordination system of ninety degrees in relation to each other is formed. The middle element generates connections among all the pieces, and it creates the force that keeps them together.

The middle becomes not just a mechanical fact; it is also a metaphor for the power of the Cube in the world. Just as the secret cross in the middle holds in place all the cubies, so, too, in its quiet, stable way, the Cube brings people together. And no matter how frustrating, no matter how educational, this three-dimensional puzzle essentially invites everyone to pause and start to play. And once they do, they have the expression on their faces I saw long ago in the park in Budapest with my daughter, with the disheveled little boy and the prosperous young mother. They are faces in repose but also intently engaged. Concentrating, turning inward, losing links with their surroundings and the external world. They look as if they are in a state of meditation.

A FEW YEARS AGO I was in Paris for a speedcubing world championship. I have grown accustomed to these events—the intensity of the young people, the remarkable speed with which they arrive at the solution, the thrill of competition that breaks down to the tiniest fractions of seconds. (Just to give you some idea, a few years later, I witnessed a pair of twins competing and the victor won by only one millisecond, *one thousandth of a second*.) When I attend these events, I am now the éminence grise, the distinguished old creator, whom many esteem and others find perplexing. How could this unassuming, rather unimpressive elderly Hungarian gentleman have possibly created this miraculous object? But they crowd around me, holding Cubes (many of them counterfeit), creased papers, notebooks, photos, and T-shirts—asking me to sign them as if I were some sort of famous actor, athlete, or politician.

At the Paris games, I had retreated from the crowds at the exhibition hall, into a small room adjacent to the stage where those who were working at the competition could take a break. While I was there, I was told that two children were looking for me and would it be okay to let them visit. Of course, I said. A girl of about eleven and her younger brother, who must have been around six, appeared, each holding a Cube they wanted me to autograph. The girl told me her little brother could not solve it yet, but that she had started a year before and was getting faster and faster. I asked her how fast, and she looked shy when she admitted that her time was about eighteen seconds—which is very impressive, and is much better than my own time, but in this world of speedcubers, it isn't much, and she knew it.

"I love it very much," she said, as if she were talking about a friend, and I realized that she was.

I signed their Cubes. Then it was time for them to leave. They had just started to walk away when the little boy turned around, ran up to me, and hugged me with real warmth and spontaneity and innocence. I was taken aback and found myself surprisingly moved by the moment. He looked up and smiled, then raced away to join his sister.

It was all so quick. I remained standing, thinking about the two children, and about how many others this object has touched. When I made it, and it launched into the world, I had faith in the Cube. I knew it was interesting, even exciting. Deep down I believed it would reach people. Even though I have always believed that intellectual curiosity was part of our common humanity, I had to struggle to keep this faith alive against the accumulating evidence to the contrary. I was confident that those whose way of thinking was close to mine—architects, designers, engineers—would be interested. Mathematicians and scientists might also find the theoretical problems involved in its solution compelling. And people who like puzzles and children who love playing would also be drawn to it.

Still, I was surprised to see that its impact was not confined to these sections of society only—that it found its way to people whom nobody would ever have thought might be attracted to it. Whether in developed or developing countries, large capitals or small villages, in the countryside or at tourist attractions, museums or art galleries, you would watch with a kind of wonder at the intensity of this engagement. The Cube has reinforced my conviction that there is such a thing as universal human nature that has nothing to do with age or status or race, nor does it have anything to do with where we were born or on what or where or how we live.

For all of its many powers, its most significant and maybe the key to unlocking one of its deepest mysteries is the connections it makes. What is hidden in the Cube can be found in each of us: the capacity to be independent and connected all at the same time. The capacity to still feel a childlike sense of discovery, of wonder, of a kind of innocent pride, no matter how old, or even jaded, we may be.

AFTER LONG, HARD WORK alone, or with the assistance of a YouTube video or a solution booklet, you may have a finished Cube. You can show the world you have done it! But then, why do you feel it is still not finished, when the task is done? Somehow, there is some urge that makes you feel that you need to do it again. Maybe it was just an accidental success. Or maybe you can go further, not just in solving it, but something else. Discover more. Learn more. Understand more fully.

Usually once a puzzle has been solved, it is done. The last piece is put in place in the jigsaw; the picture is ready to be framed or discarded.

Not so with the Cube. And that is one of its most mysterious qualities. The end turns into new beginnings.

INTERVIEW WITH THE AUTHORS

In my dream, I was two cats, playing with each other.
—FRIGYES KARINTHY

Three should be even better.
—THE CUBE

Questioner: *Thank you very much for joining us, Mr. Rubik. And before we begin, I have a Cube with me, and could you possibly autograph it? Thank you! Okay, let's start with a simple question: Are you satisfied with your book?*

(In unison)

Rubik: As usual, I had higher expectations.

Cube: Yes, absolutely!

Q: *Why don't you each explain?*

R.: As I already said at the beginning of the book, I have problems with writing and verbal expression. I wanted a book that captured the Cube's mysterious existence in the world as well as my life beyond him. When I began this project, I was determined that the book not have an evident structure. Or a narrative. I definitely didn't want chapters and imagined it to have almost no beginning and no end. Of course, that is not the way things work out in the real world. My hope now is to leave it to the readers to decide what they think about its structure, and I hope that they are smarter than me.

C.: As for me, I am extremely pleased! Even now, in my forties, my very rich and interesting life has only just begun, but no one has ever asked yet for my side of the story. So finally I can tell it! I love to speak!

Q: *Why is* Cubed *the title of this book?*

R.: The best title for me would have been *no* title. But, as we all know, titles are considered to be necessary. Since the Cube's name is linked to mine, and since I am the author, it seemed only fair to share the title with him. When the Cube was born, I originally named it Magic Cube. But there is bad and good magic. With bad magic, someone with great powers turns the boy into a frog; but then, with good magic, the frog can be turned into a prince. I was under the Cube's spell and named my offspring almost in a protective way—like when the king adopts a child who could not be in the formal line of succession. The first time I named the Cube, it was emotional. But then it became a rational and legal question.

We needed a title. I wanted it to be connected with the Cube. Not directly, but with a twist. You probably know, the term "cubed" is related to the Cube as a geometrical form, and speaks about volume, which only exists in space. In one dimension, we measure distances; two dimensions, we have only area; and in three dimensions, there is volume. To be honest, Cube, the expression is more about you than me. But in the end, I hope you realize that it is much more about the people who love you.

C.: I hardly understood what you just said, and find myself lost in the middle. But what could possibly be wrong with that?

Q: *How do you both feel after all this?*

(In unison)

C.: Excited! More alive! Curious! Playful! I want to begin all over again.

R.: Exhausted. Dying. Bored. More seriously, I feel as if I made something that I know is not really finished yet, because now starts the hard work. As you know, I am an architect; so, for me, the book you are holding is just the draft model. Now it's time for people to read it, and that means teamwork and sharing. Unfortunately, all the modifications will have to wait.

Q: *You are both very different. How would you describe the contrasts and contradictions between you two?*

R.: That's easy, and you can see it in this conversation. I don't like to speak about myself and the Cube loves speaking about himself. In fact, he loves that we are both speaking about him. The Cube is very well socialized; I am less so. The Cube doesn't need to know languages; he can communicate with everybody, though not verbally. I know only Hungarian and some English. He is perfect. I am looking for excellence only, and that is rarely achieved. The Cube never changes his appearance. But I have—gray hair, wrinkles, glasses for reading. The Cube is immortal, and I am not.

C.: That sounds so sad! For me, I live not in the past, not in the future, but in the present moment!

R.: As you may have already read, the only problem is that the "present" doesn't exist. It is only a moving dot with no dimension on the timeline, coming from the past and going to the future infinitely. But you can say the opposite, too, that only present *exists*; past is over and future not born yet.

Q: *May I interrupt you? I have a question for the Cube: How does it feel to be the subject of a book?*

C.: I love it! But look, I have been already the subject of so many books, it isn't really new for me. What *is* new, and I think rather interesting, is that this we are doing together. These movements are very familiar to me. I have a sense of turning. And it is very special for me to see him working on me in a new way. He started it nearly fifty years ago, but now he has a different opportunity and a different approach. I can't say it is better. In fact, it may be worse. Now that I think of it, I should have had many more opportunities to express myself.

R.: I know that you didn't ask me this question, but there's something I would like to add. I have a comment, not on the answer—but on the question. This is a good example of how a question can be a statement of fact at the same time. That is not necessarily a problem in itself; it's only a problem if the statement is not true. He is not the subject of this book. You can say he is one of the subjects, more important or more interesting and colorful than I, as a subject. The real subject is *we*; it is all of us.

Q: *So, what are your plans for the future?*

C.: I would like to be involved more and more in education. In sports, my goal is the Olympics! And I think it would be nice to have an art gallery. I am already a movie star. I have often been the supporting actor, but I am looking forward to the leading role. There are monuments in my honor, but I am still alive. But I don't want to talk about a cable TV channel because, as you know, I am very modest.

R.: I will agree with him that working more in education would be a great thing. I also think that there may be a few more ideas left in me to bring more life into my endless retirement.

Q: *The last question: Could you describe what you mean to each other?*

R.: Hmmm . . . I think we are close. We are partners. We are coconspirators. We understand each other's essential nature in a very deep way. And in the end, of course, the Cube will always be the star, and that is just fine with me.

C.: Of course, he's one of my parents next to Mother Nature, and I believe in him! I'm his creation, after all. But somehow, I changed him, too. I opened up his life the way he opened up mine—without me, he would be just another Hungarian fellow with some crazy ideas.

Oh my, look at the time! Sorry, guys, I have another appointment with some of my fans. I must leave you now . . . but why don't you two continue to play?!

PUBLISHER'S NOTE

It has been a privilege to work with the author in bringing this book to readers, from Hungary to all over the world. We are tremendously grateful to Marianne Szegedy-Maszak, whose gentle touch is palpable on every single page, as well as to András Forgách and Márk Baczoni, for their support and assistance throughout this process. Our thanks goes also to our colleagues in the United States, especially to editor Bryn Clark and Bob Miller, president of Flatiron Books, who helped us make this book happen. *Cubed* owes most of all to the people who ever played with a Cube.

—Adam Halmos,
Libri Publishing, Budapest

ABOUT THE AUTHOR

Ernő Rubik is a professor of architecture and a creator of mechanical puzzles, most notably the Rubik's Cube. He has lived in Budapest, Hungary, all his life.